MISSION IMP

Patrick Bury is a lecturer in Defence and Strategic Studies at the University of Bath; his research focuses on military and counter-terrorism organisational transformation and cohesion. Prior to entering academia, Patrick served in the British Army as an air assault infantry combat officer, fighting in one of the most dangerous areas of Afghanistan: Sangin. A memoir of his platoon's tour, *Callsign Hades* was published in 2010 and has been described as 'the first great book of the Afghan war'. After leaving the Army, Patrick worked as an analyst for NATO and then a private security firm. He has 15 years' experience of working in the security sector as practitioner, analyst and academic.

To the men and women of the Army Reserve

Mission Improbable

The Transformation of the British Army Reserve

PATRICK BURY

Howgate Publishing Limited

First published in 2019 by
Howgate Publishing Limited
Station House
50 North Street
Havant
Hampshire
PO9 1QU
Email: info@howgatepublishing.com
Web: www.howgatepublishing.com

British Library Cataloguing-in-Publication Data
A catalogue record for this book is available from the British Library

ISBN 978-1-912440-04-7 (pbk)
ISBN 978-1-912440-05-4 (ebk - PDF)
ISBN 978-1-912440-12-2 (ebk - ePUB)

Patrick Bury has asserted his right under the Copyright, Designs and Patents Act, 1988, to be identified as the author of this work.

Contents

List of Figures

Acknowledgements

This book would not have been possible without the assistance and guidance of numerous academics, military personnel and units that gave generously of their time. I am profoundly thankful to Professor Tony King for his close guidance, insightful analysis, and expertise in the development and refinement of the PhD thesis upon which this book is based. Professors Tim Edmunds, Vince Connelly, Caroline Kennedy-Pipe and David Galbreath, and Dr Sergio Catignani gave excellent and detailed feedback on how to improve the text for which I am also very grateful. Thanks also to Dr Alex Neads, Dr Patrick Finnegan, and Dr James I. Rogers for comments on chapters and insights into reserve service. Kirstin Howgate at Howgate Publishing believed in the project and assisted with the editing and production throughout. Similarly, the project would not have been possible without an Economic Social and Research Council studentship kindly awarded to me by the University of Exeter. I would also like to thank the postgraduate team at Exeter for their kind support during the PhD process. I am also indebted to the British Army for allowing access to their soldiers and for funding some of the research. In terms of the participants, I am very grateful to the numerous generals, ministers, and senior officers who gave generously of their time. Most of all, I would like to express my sincere gratitude to the men and women of the Army Reserve, in particular those in RLC and REME units, who participated in the study. This book could not have been written without your experiences and candid insights, so thank you for taking the time to share them.

List of Abbreviations

AGAI	Army General Administrative Instruction
AR	Army Reserve
BEF	British Expeditionary Force
CDS	Chief of the Defence Staff
CGS	Chief of the General Staff
COB	Combat Operating Base
Coy	Company
DLTP	Defence Logistics Transformation Programme
DoD	U.S. Department of Defense
EST	External Scrutiny Team
Fd Coy	Field Company
FFMA	Forward Force Maintenance Area
FMA	Forward Maintenance Area
FOB	Forward Operating Base
FOC	Full Operational Capability
FR20	Reserves in the Future Force 2020 Policy
IED	Improvised Explosive Device
IOC	Initial Operating Capability
JIT	Just-in-Time
LAD	Light Aid Detachment
LSD	Logistics Support Detachment
MATT	Military Annual Training Test
MJDI	Management of Joint Deployed Inventory
MoD	Ministry of Defence
MPA	Major Projects Authority
MTDs	Man Training Days
NAO	National Audit Office
NSPA	NATO Support and Procurement Agency
NSS	National Security Strategy
PSI	Permanent Staff Instructors
REME	Corps of the Royal Electrical and Mechanical Engineers
ResCAS	Reserve Continuous Attitudes Survey
RLC	Royal Logistics Corps

RMA	Revolution in Military Affairs
RML	Revolution in Military Logistics
SCM	Supply Chain Management
SCOR	Supply Chain Operations Reference model
SDSR	Strategic Defence and Security Review
Sqn	Squadron
TA	Territorial Army
TAV-	Total Asset Visibility Minus
TSF	Total Support Force
TRANSCOM	Transport Command

Chapter 1

The Rise of the Reserves

On 3 July 2013 in the House of Commons, the then Defence Secretary Philip Hammond outlined perhaps the most radical transformation of the Territorial Army (TA) attempted since its inception 105 years before. Summarising the new *Reserves in the Future Force 2020: Valuable and Valued* (FR20) policy, Hammond announced that in order to arrest the decline of the reserves and better integrate them with the regular armed forces, the government was investing £1.8 billion over the next ten years in reserve equipment, training and remuneration.[1] £1.2 billion of this investment would focus on the TA – by far the largest of Britain's four reserve forces – to increase both its size and military capability. The *quid pro quo* of this investment was that the reserves would increase their military capability, become much more closely integrated with the regulars, and deploy more often. As Hammond outlined: 'The job that we are asking our reservists to do is changing, and the way in which we organise and train them will also have to change', while FR20 itself went on to state that: 'We will use our Reserve Forces to provide military capability as a matter of routine, mobilising them when appropriate.'[2] Decisively, FR20 placed major emphasis on outsourcing military logistics capability previously held in the regular army to an expanded and more deployable reserve logistics component. Crucially, as explored in this book, FR20 outlined significant changes to the capabilities expected of reserve logistics sub-units, stating: 'Greater reliance will be placed on the Reserves to provide routine capability ... primarily in the areas of combat support (artillery and engineers), [and] combat service support (such as logistics, medical).'[3] Such a transformation envisaged the centralisation of reserve units and their incorporation into the army's new tiered readiness structure, 'Army2020'. This new vision articulated a step-change in the prominence of the reserve army in British defence policy and a major transformation of a force that had traditionally been a part-time militia of citizen-soldiers. The challenge was great, but with Hammond stressing the investments to be made to the reserves in numerous areas, FR20 received wide cross-party support in the House that day.

1   Ministry of Defence (2013) *Reserves in the Future Force 2020: Valuable and Valued* (White Paper), ondon: HMSO. Henceforth *Future Reserves 2020*.
2   *Hansard* (2013) The Secretary of State for Defence (Mr Philip Hammond) Reserve Forces: Statement to House of Commons, 3 July, col. 924; *Future Reserves 2020*, 12.
3   *Future Reserves 2020*, 22.

The attempt to transform Britain's reserve army from a strategic to an operational reserve had begun, and FR20 would quickly become a central tenet of British defence policy in the Cameron era.

The support Hammond received unveiling FR20 in the Commons in July 2013 stood in stark contrast to its genesis, and indeed, its later evolution. Before the government had even unveiled the transformation, it had had to reconcile intra-party political divisions, overcome resistance from army high command, then set up a separate planning team due to lingering distrust, while all the time remaining sensitive to public opinion in the wake of the recent cuts to the defence budget and the size of the army in particular. Indeed, in a nod to the impact these cuts had had on the army, Hammond remarked: 'The Army ... has had substantially to redesign its reserve component to ensure that regular and reserve capabilities seamlessly complement each other in an integrated structure designed for [its] future role.'[4] Hammond thus highlighted that the transformation of the reserves was closely related to, and was being undertaken simultaneously with, what was arguably the most significant organisational transformation of the army since the abolition of conscription in 1960.

Driven by political, financial – and to a much lesser extent – strategic imperatives, the National Security Strategy and Strategic Defence and Security Review (SDSR) of 2010 signalled the new Conservative–Liberal Democrat coalition government's primary desire to prioritise the economic security of the United Kingdom in the wake of the 2008 global recession.[5] However, it also represented a political desire to avoid the long-term interventions of the Afghanistan and Iraq wars. This had dominated the British public's perception of the armed forces during the same period, left them questioning previous governments' decision-making, and exposed major tensions between senior military commanders and their political masters.[6] Nevertheless, the Cameron government's desire to reduce defence spending forced significant changes on the army, including how it perceived its future operations and how it organisationally oriented itself toward fulfilling them. The resulting transformation, labelled *Army2020*, adopted a contingency-based approach to operations and emphasised defence engagement as one of its core tasks.[7] However, the most profound element was the reduction in regular army manpower from

---

4   *Hansard*, 3 July 2013, col. 924.

5   HM Government (2015) *National Security Strategy and Strategic Defence and Security Review 2015: A Secure and Prosperous United Kingdom*, Norwich: HMSO; Cornish, P. and Dorman, A. (2011) 'Dr Fox and the Philosopher's Stone: the alchemy of national defence in the age of austerity', *International Affairs*, 87(2).

6   Kellner, P. (2012) 'Public Perceptions of the Army', RUSI Land Warfare Conference, 3 July 2013, available at https://www.youtube.com/watch?v=eS6DgD5ZUSM; Bailey, J., Iron, R. and Strachan, H. (eds) (2013) *British Generals in Blair's Wars*, London: Routledge; Strachan, H. (2003) 'The Civil-Military Gap in Britain', *Journal of Strategic Studies*, 26(2).

7   Ministry of Defence (2013) *Transforming the British Army, An Update – July 2013*, London: Ministry of Defence, 1.

102,000 in 2010 to 82,000 by 2018.[8] This reduction in personnel resulted in a new structure and readiness model for the army, and in particular, a renewed emphasis on the integration of the TA (soon to be re-christened the Army Reserve) to support the readiness cycle. Thus, the reductions in regular personnel were to be offset by a larger and more deployable force of army reservists, whose combined Phase 1 and 2 (fully trained) trained strength was to be expanded from 19,230 to 30,000 by 2018.[9] Much of this expansion was focused on the logistics component, which was expected to now routinely provide the logistics capability stripped from the regulars. On paper at least – and certainly, as will be discussed, it was presented in this manner by the government – FR20 was therefore central to the success of Army2020. In the following months and years, this repositioning of the Army Reserve at the core of British defence policy would ensure strong political and media interest in its evolution, and heavy criticism of its failures. But what exactly did FR20 aim to achieve?

FR20 represented the most severe transformation of the army's reserve since the Haldane reforms of 1907–1908 created the TA, linked it with the regular army's regimental system, and ensured that TA units would be raised locally.[10] Most decisively, the full integration of the new Army Reserve into the Army2020 force readiness structure represented a fundamental change to the organisation's once peripheral place in British defence policy. The traditional evolution of the TA, bureaucratic politics and the 1996 Reserve Forces Act, limited the deployability of the TA and the roles it fulfilled, especially abroad.[11] In essence, FR20 envisaged the Army Reserve as more highly trained, more deployable, and therefore more capable of operating with their regular counterparts. Crucially, it stated:

> Under our new model, the use of the Reserves is no longer exceptional or limited to times of imminent national danger or disaster, but is integral to delivering military effect in almost all situations ... As an integral part of the Armed Forces, reservists will be required for almost all military operations ... [and] principally in the Army's case and as the situation demands, as formed sub-units or units.[12]

---

8   Ministry of Defence (2012) *Transforming the British Army, July 2012: modernising to face an unpredictable future*, London: Ministry of Defence. This number has been achieved ahead of schedule, see 'British Army already below smaller 82,000 target', *The Daily Telegraph*, 29 July 2015.

9   *Future Reserves 2020*, 11; Ministry of Defence (2013) *UK Reserve Forces and Cadets 1 April 2013*, London: MoD, 6.

10   Territorial and Reserve Forces Act 1907. It actually created the Territorial Force, the Territorial Army and Militia Act 1921 changed the name from TF to TA.

11   Ministry of Defence (2011) *Future Reserves 2020 – The Independent Commission to Review the United Kingdom's Reserve Forces*, London: HMSO, 14–18. Henceforth *The Independent Commission*.

12   *Future Reserves 2020*, 7, 17.

As such, FR20 aimed to change the traditional perception of the TA as a part-time force for use only in time of great emergency; the Army Reserve will now deploy routinely and aims to potentially compel employers to release personnel through changed legislation.[13] Similarly, it outlined a change in the nature of how reservists are to be used on operations. Taken together, this transformation marked a step-change in the liability for the Army Reserve and in its role from a strategic to an operational reserve. FR20 also detailed the closure and centralisation of a number of local Army Reserve sites in order to increase efficiencies during peacetime. Thus, FR20 aimed to transform the structure, role and capabilities of the Army Reserve.

The Post-Fordist Rise of the Reserves

With its conscious emulation of American, Australian and Canadian reserve forces, FR20 is reflective of international developments concerning the use and reorganisation of these forces. However, while international security scholars have recognised the importance of military professionalisation,[14] transformation,[15] and civil–military relations[16] in shaping and understanding security outcomes, these debates have almost entirely been in relation to regular combat forces. Yet, since 2001 there has been a marked increase in the use of reserve forces in conflicts, with the US deploying an entire reserve division during the invasion of Iraq in 2003 and thoroughly increasing its reserve capability through a sustained transformation process thereafter.[17] At their peak, reservists contributed 20 per cent of British Army

---

13 *Future Reserves 2020*, 9.

14 Huntington, S.P. (1957), *The Soldier and the State: The Theory and Politics of Civil–Military Relations*, Cambridge, MA: Harvard University Press; Janowitz, M. (1971), *The Professional Soldier: A Social and Political Portrait*, New York: Free Press; King, A. (2013), *The Combat Soldier: Infantry Tactics and Cohesion in the Twentieth and Twenty-First Centuries*, Oxford: Oxford University Press.

15 Cohen, E. (2004), 'Change and Transformation in Military Affairs', *Journal of Strategic Studies* 27(3), 395–407; Biddle, S. (2004), *Military Power: Explaining Victory and Defeat in Modern Battle*, Princeton: Princeton University Press; King, A. (2011), *The Transformation of Europe's Armed Forces*, Cambridge: Cambridge University Press; Farrell, T., Rynning, S. and Terriff, T. (2013) *Transforming Military Power since the Cold War: Britain, France, and the United States, 1991–2012*, Cambridge: Cambridge University Press.

16 Huntington, *The Soldier and the State*; Edmunds, T. (2012), 'British civil–military relations and the problem of risk', *International Affairs*, 88(2), 265–82; Bailey, Iron and Strachan (eds), *British Generals*.

17 Schnaubelt, C. et al. (2017), 'Sustaining the Army's Reserve Components as an Operational Force', Santa Monica: RAND Corporation, available at https://www.rand.org/pubs/research_reports/RR1495.html.

manpower to operations in Iraq and 12 per cent in Afghanistan.[18] More recently, the resurgence of the Russian hybrid threat has caused numerous European militaries to reform their reserve forces. Ukraine called up 100,000 conscripts into its reserve in 2014 and is currently reorganising its reserve.[19] In 2015, Poland announced it was creating a 50,000-strong reserve Territorial Defence Force.[20] The same year, Finland put its 900,000 reservists on notice for mobilisation and intensified its reserve transformation, while 2018 saw Sweden mobilise its 22,000-strong Home Guard for the first time since 1975.[21] Norway has recently increased its reserve forces too. Meanwhile, British reservists continue to contribute to NATO's 'trip-wire' presence in Estonia. Further afield, Canada, Australia and Argentina are currently transforming their reserve forces.[22] In the current era therefore, reserve forces matter.

It is within this context of an international rise of reserve forces that this book sits. These recent transformations and use of reserves clearly highlight that they are an increasingly important aspect of international security. Crucially, they offer the potential of cheap, scalable mass to military planners operating under increasingly tight fiscal constraints by outsourcing military capability to reserve forces. Nevertheless, if they are to deliver effectiveness and efficiencies, training and force structures must be robust and realistic if reservists are to be capable enough to rapidly respond to threats. Indeed, at the heart of many of these transformations lies the uneasy dichotomy between the professionalisation of most Western armies since the 1960s through the interrelated processes usefully described as 'post-Fordist', and the increasing reliance on part-time citizen-soldiers who are now to be better integrated with their full-time professional counterparts. Initially, following the post-modernist trend, numerous scholars have examined how societal changes and increasing post-Cold War strategic uncertainty has resulted in changes to the missions and structures of the modern military.[23] For James Burk and Charles Moskos, recent

---

18 Feur, A. (2005), 'Regular Citizens With Regular Army', *New York Times*, 6 February 2005; HM Government (2011), Ministry of Defence (2011) *The Independent Commission*, 11.

19 Akimenko, V. (2018), 'Ukraine's Toughest Fight: The Challenge of Military Reform' Carnegie Endowment for International Peace, 22 February, available at https://carnegieendowment. org/2018/02/22/ukraine-s-toughest-fight-challenge-of-military-reform-pub-75609

20 Gao, C. (2018), 'This Is How Poland Plans to Fight Russia in a War'. *The National Interest*, 3 March, available at http://nationalinterest.org/blog/the-buzz/how-poland-plans-fight-russia-war-24731.

21 Finnish Military, available at http://www.ft.dk/samling/20151/almdel/fou/spm/321/ svar/1345997/1670318.pdf; https://www.thelocal.se/20180606/sweden-mobilises-entire-home-guard-for-first-time-since-1975, both accessed 22 May 2018.

22 See Lt Gen. Angus Campbell (2017), http://dra.org.au/conference-2017-item/27405/ australian-army-reserve-transformation-a-total-force/?type_fr=813 and http://www.forces. gc.ca/en/about-reports-pubs/transformation-report-2011.page both accessed 13 June 2018.

23 Baudrillard, J. (1994), *Simulacra and Simulation*, Michigan: University of Michigan Press; Lyotard, J-F. (1979), *The Postmodern Condition: A Report on Knowledge*, Manchester: Manchester University Press.

societal evolutions have resulted in changing conceptions of the rights and duties of citizens with regards to military service, while simultaneously underpinning an organisational shift from conscript forces designed to partake in mass state-on-state conflicts towards a 'smaller, voluntary professional force that relies on reserve force to accomplish its missions.'[24]

However, the post-modern view fails to address economic and industrial change, and while the debate over the extent to which modern militaries are truly post-modern continues,[25] Anthony King has developed the term of post-Fordism to describe the ongoing changes in Western militaries.[26] King draws on industrial sociology to examine how the end of the Fordist mode of production, relying on mass labour forces 'employed on long-term contracts, producing standardised products for stable markets' began to be undermined in the 1970s by rising production costs and competition.[27] In response to these dual pressures, companies in Japan and America began to organisationally transform. Four central changes have been identified in this transformation: the replacement of mass labour with a highly skilled core and less-skilled periphery; the outsourcing of non-core functions and the adoption of 'just-in-time' (JIT) logistics to reduce overheads; the centralisation of headquarters and the flattening of industrial hierarchies; and the development of a network approach to supply and knowledge.[28]

For King, the professionalisation of Western militaries, their continued reduction in size, and the concentration of military power in the special forces, are indicative of the development of a highly specialised core, while the increasing emphasis on surging reserve manpower in times of need highlights the periphery.[29] The US military's outsourcing of specialist logistical and technical services is briefly discussed while King also acknowledges the adoption of JIT logistics practices to reduce overheads. Centralisation is evident in the development of joint and transnational military headquarters which share professional knowledge while paradoxically encouraging subordinates to act on their own initiative by decentralising command decisions, thereby flattening hierarchies. Similarly,

---

24 Moskos, C. and Burk, J. (1994) 'The Post-Modern Military', in Burk, J. (ed.), *The Military in New Times: Adapting Armed Forces to a Turbulent World*, Boulder, CO: Westview Press, 171.

25 Dandeker, C., Booth, B., Kestnbaum, M. and Segal, D. (2001) 'Are Post-Cold War Militaries Postmodern?', *Armed Forces and Society* 27(3); King, A. (2005) 'Towards a Transnational Europe: the case of the armed forces', *European Journal of Social Theory*, 8(4); Bondy, W. (2001) 'Postmodernism and the Source of Military Strength in the Anglo-West', *Armed Forces and Society*, 31(1).

26 Dandeker, et al., 'Are Post-Cold War Militaries Postmodern?'; Kaldor, M., Albrecht, U. and Schméder, G. (eds), (1998) *The End of Military Fordism*, London: Pinter, 2.

27 King, A. (2006a) 'The Post-Fordist Military', *Journal of Political and Military Sociology*, 34(2), 360.

28 Ibid., 360–61.

29 Ibid., 361–64.

the development of a non-linear operational approach to the dispersal and co-coordination of forces centred around independent brigades indicates the military's adoption of a network approach to warfare.[30] Using this evidence, King argues that modern Western militaries have transformed in a fashion analogous with post-Fordist industry, primarily due to similar 'supply and demand-side pressures.'[31] He draws on the wider literature on institutional transformation to posit that, faced with these pressures, Western militaries have emulated industry in a process similar to the 'institutional mimetic isomorphism' first coined by Paul Dimaggio and Walter Powell.[32]

King's contribution is an accurate description of the changes occurring within Western militaries and is perceptive as to why these are happening. In identifying dominant modes of production, and economics, as important sources of military transformation, his approach explicitly links military change with industrial and economic change. However, King's main focus remains on land combat forces. While he notes the role of logistics in wider military transformation, in particular in relation to outsourcing, the exact nature and impact of these logistical changes is not fully developed. The question remains if logistics in modern militaries – presumably under similar, if not more intense, economic pressures than the combat function – have transformed in a similar post-Fordist fashion.

Crucially in terms of my argument, in the post-Fordist mode of production the distinction is made between the 'specialist core ... and a subsidiary workforce on temporary and short-term contracts.'[33] Clearly, these core/periphery observations have immediate relevance for the current transformation of the British Army, with Army2020 reorganising the force into a Reaction/Adaptable Force structure, and its renewed emphasis on the reserves. Moreover, in keeping with the reasons for these changes, and countering the fluidity associated with post-modernism, King argues that militaries are 'changing in structure to fulfil new missions in the face of economic and strategic pressures.'[34] Similarly, the outsourcing of defence tasks to the reserves and the tiered readiness outlined in Army2020 are indicative of the post-Fordist trend toward JIT delivery of services and supplies to increase efficiency. As such, in examining the current changes to the structure and role of the reserves, and in particular their logistics component, the post-Fordist literature provides the overarching conceptual framework for understanding why and how Britain is attempting to transform the Army Reserves' effectiveness, especially in terms of logistics.

---

30 Ibid., 367.
31 Ibid., 368.
32 Dimaggio, P. and Powell, W. (1983), 'The Iron Cage Revisited: Institutional isomorphism and collective rationality in organisational fields', *American Sociological Review*, 48(2).
33 King, 'The Post-Fordist Military', 360.
34 Ibid., 362.

Military Transformations

Another useful way of looking at FR20 is as an organisational transformation. That FR20 is an attempt to transform the reserves is clear. The FR20 policy document itself stated that 'FR20 is part of the wider Transforming Defence campaign that is aiming to transform our Armed Forces and deliver Future Force 2020.'[35] It also specifically mentioned reserve transformation a further three times, placed it within the context of the Army2020 transformation, and made the 2-star Director General Army Transformation responsible for implementing the policy.[36] That the army and the wider defence establishment viewed Army2020 and FR20 as a transformative process is also clearly supported by other official documents.[37] Conversely, reform is not mentioned once in relation to the reserves in FR20.[38]

Recently a significant body of literature has emerged that considers the sources of transformation within military organisations. Theo Farrell and Terry Terriff have defined military transformation as a major 'change in the goals, actual strategies, and/or structures of a military organisation.'[39] Crucially, they argued that 'it is the outcome of military change that determines whether it is major or minor in character.'[40] Broadly speaking, two main schools of thought have developed on how militaries change. The top-down approach of Barry Posen, Steven Rosen, Deborah Avant and Kimberley Zisk, focuses on the importance of doctrine, civil–military relations and inter- and intra-service politics as drivers of military transformation.[41] In particular, Posen identified how civilian leaders identify and promote 'maverick' officers who agree with their vision when instigating transformation, while Rosen argued that in fact intra-service rivalry over future visions of victory drove innovation. Interestingly for this study, Zisk later refuted Rosen's conceptualisation of the services as too monolithic and instead showed how innovation is a more complex process of alliance building between interest groups within organisations. These top-down transformations – what Farrell

---

35 *Future Reserves 2020*, 59.

36 Ibid.

37 Ministry of Defence, *Transforming*; Ministry of Defence, *Transforming – An Update*.

38 *Future Reserves 2020*; Ministry of Defence, *Transforming*; Ministry of Defence, *Transforming – An Update*.

39 Farrell, T. and Terriff, T. (eds) (2001) *The Sources of Military Change: Culture, Politics, Technology*, Boulder, CO: Lynne Rienner, 6; Farrell, T. (2010) 'Improving in War: Military Adaptation and the British in Helmand Province, Afghanistan, 2006–2009', *Journal of Strategic Studies*, 33(4), 570.

40 Farrell and Terriff, *The Sources of Military Change*, 5–6.

41 Posen, B. (1984), *The Sources of Military Doctrine: France, Britain and Germany Between the World Wars*, New York: Cornell University Press; Avant, D. (1994), *Political Institutions and Military Change: Lessons from Peripheral Wars*, New York: Cornell University Press; Rosen, S. (1991), *Winning the Next War: Innovation and the Modern Military*, New York: Cornell University Press.

later labelled 'innovation' – represents most of the previous attempts initiated by political or military elites to reform British reserve forces in the past.[42] As Chapter 2 highlights, major reserve reform has traditionally been a top-down process. However, more recently Adam Grissom, Eliot Cohen, and James Russell, amongst others, have argued that militaries can also transform in response to bottom-up – or tactical – pressures.[43] Grissom has argued that bottom-up tactical changes can be simultaneously involved in transformation,[44] and Farrell later conceptualised these processes as top-down innovation – a 'major change that is institutionalised in new doctrine, a new organisational structure and/or new technology' – and bottom-up 'adaptation' which represents a 'change to tactics, techniques or existing technologies to improve operational performance'.[45] Rob Foley, Stuart Griffin and Helen McCartney have shown how top-down and bottom-up changes are largely dependent on each other if transformation is to be lasting.[46]

Clearly then, there are different approaches to understanding military transformations, and scholars have recently begun to acknowledge the complexity of transformative processes. Very recently, Stuart Griffin has excellently critiqued the transformation literature. While lauding the discipline for its open, multidisciplinary approach, he argues that it has predominantly followed the cultural turn. Decisively, he argues that it frequently lacks the sustained application of wider organisational and sociological theory.[47] I seek to address this lack of broader theoretical inquiry in transformation studies by incorporating not only the post-Fordist conceptual framework for explaining organisational change, but also the sociological literature on professionalism and cohesion to give greater theoretical depth to my evidence on the nature of the Army Reserve and my arguments on FR20 as an attempt to transform it.

While different aspects of the transformation literature run through this study, I primarily draw on two major contributions. The first is the top-down innovation literature detailed above, which is particularly pertinent as it provides the closest conceptual link between the transformation literature and previous works on the British reserves. The second concerns normative transformative patterns. Elizabeth Kier has challenged the top-down approach's realist-functional focus, arguing that organisational culture, rather than institutional politics and power, explains

---

42 Farrell and Terriff, *The Sources of Military Change*, 6; Farrell, T. 'Improving in War: Military Adaptation and the British in Helmand Province, Afghanistan, 2006–2009', 570.

43 Grissom, A. (2006), 'The Future of Military Innovation Studies', *Journal of Strategic Studies*, 29(5), 910.

44 Russell, J. (2010), 'Innovation in War: Counterinsurgency Operations in Anbar and Ninewa Provinces, Iraq, 2005–2007', *Journal of Strategic Studies*, 33(4).

45 Farrell and Terriff, *The Sources of Military Change*, 6; Farrell, T. 'Improving in War', 570.

46 Foley, R., Griffin, S. and McCartney, H. (2011), '"Transformation in Contact": learning the lessons of modern war', *International Affairs*, 87(2).

47 Griffin, S. (2017), 'Military Innovation Studies: Multidisciplinary or Lacking Discipline?', *Journal of Strategic Studies*, 40(1/2), 15–16.

the choice of offensive and defensive doctrinal postures. For Kier, doctrine 'is best understood from a cultural perspective'.[48] She supports her arguments with evidence from the inter-war years of the British military's refusal to professionalise due to concerns about control of the military inherently bound in British history and culture, and with evidence showing that competing ideologies on the political Left and Right in France about the military's role in society curtailed its ability to increase its effectiveness, resulting in a defensive doctrinal posture. Kier drew heavily on Ann Swidler's definitions of culture and ideology, which is worthy of repetition here. For Swidler, culture is defined as 'the set of assumptions so unself-conscious as to seem a natural, transparent, undeniable part of the structure of the world', while ideology is the 'highly articulated, self-conscious belief in [a] ritual system aspiring to offer a unified answer to the problems of social action.'[49] Thus, culture can be perceived of as an inherent cause of action, ideology an explicit call for a certain kind of action. However, while adding a rich cultural perspective, Kier's analysis is also top-down, doctrinal-based, and focused solely on regular combat forces. Building on Kier's work, Farrell also used a constructivist approach to highlight the importance of cultural norms within military organisations in relation to change.[50] Farrell and Kier were right to identify the importance of culture in influencing transformations. However, Farrell's analysis is predominantly concerned with militaries' tendency to emulate others' organisational structure and doctrine, and although Kier discusses professionalism in the context of the British military, she does not examine in detail the impact that professional culture can have on a force.

Interestingly, but unsurprisingly, all these approaches to military transformation have focused exclusively on the combat arms and how the way they conduct operations over time has changed. Similarly, King's work on the transformation of Europe's armed forces identified very important changes in the operational planning, structures and networks of combat forces exclusively.[51] King's later study of the impact of professionalisation on the modern Western soldier also focused exclusively on combat troops.[52] Indeed, none of the recent literature on military transformation has examined military logistics, nor reserve components. Crucially, it remains to be seen if and how differences in the organisational culture and bureaucratic politics of the reserves influences transformation compared to

---

48   Kier, E. (1997), *Imagining War: French and British Military Doctrine*, Princeton: Princeton University Press, 21.

49   Swidler, A. (1986), 'Culture in Action: Symbols and Strategies', *American Sociological Review* 51(2), 279, 284.

50   Farrell, T. (2001), 'Transnational Norms and Military Development: Constructing Ireland's Professional Army', *European Journal of International Relations*, 7(1).

51   King, A. (2011), *The Transformation of Europe's Armed Forces: From the Rhine to Afghanistan*, Cambridge: Cambridge University Press.

52   King, *The Combat Soldier*.

regular forces. More specifically, how is the culture of professionalism influencing the FR20 transformation of reserve logistics units? It can also be argued that the majority of the transformation literature is positivist: almost exclusively, only major transformations that have been successful have been studied. While Rosen, Avant and Kier have considered how organisational stasis and the inability to adopt the appropriate offensive or defensive military postures leave states ill-prepared for war, they do not consider transformations that have not, or have only partially, succeeded in and of themselves. Warning against this over emphasis in the literature, Griffin has called for 'revisiting some of the case studies of failure to innovate'.[53] Indeed, of the three examinations of the failure to transform, all are focused on war time transformation and do not address the issue of top-down, politically-imposed transformations in peacetime, nor consider wider sociological debates about the changed nature of modern society.[54] Overall, therefore, this leaves open the important question of why do peacetime attempts to transform reserve forces flounder?

Professionalisation

Another useful perspective on FR20 related to both post-Fordism and transformation is the literature on the professional military. There is wide consensus that the pace of Western military professionalisation vastly increased in the second half of the 20th century with the end of conscription, the reduction in armies' size and the increasing technological sophistication of warfare.[55] Samuel Huntington's *The Soldier and the State* is both dated and problematic, but it was the first to identify the changes that increasing professionalism were having on the US Army in the 1950s.[56] Huntington argued that the US professional officer corps was a 'functional group with highly specialised functions' akin to

---

53 Griffin, 'Military Innovation Studies: Multidisciplinary or Lacking Discipline?', 16.
54 Catignani, S. (2012) 'Getting COIN at the Tactical Level in Afghanistan: Re-Assessing Counter-Insurgency Adaptation in the British Army', *Journal of Strategic Studies*, 35(4); Grissom, A. (2013) 'Shoulder-to-Shoulder Fighting Different Wars: NATO Advisors and Military Adaptation in Afghan National Army, 2001–2011', in Farrell, T., Osinga, F. and Russell, J. (eds) *Military Adaptation in Afghanistan*, Stanford: Stanford University Press; Harkness, K. and Hunzeker, M. (2015) 'Military Maladaptation: Counterinsurgency and the Politics of Failure', *Journal of Strategic Studies*, 38(6).
55 Huntington, *The Soldier and the State*; Janowitz, *The Professional Soldier*; King, *The Combat Soldier*; Moskos, C., Williams, J. and Segal, D. (eds) (2000), *The Postmodern Military: Armed Forces after the Cold War*, Oxford: Oxford University Press.
56 For instance, Huntington's focus on the officer corps as the sole custodians of professional practice does not match the realities of the expertise held by professional soldiers today, while his views that most civil–military relations occur between officer and the state/public appear even more outdated given recent advances in information technology and social media.

other professions; he thus defined professionalism as a product of expertise, responsibility and corporateness.[57] He therefore noted how officers in particular were increasingly educated and trained to acquire skills and knowledge to conduct highly specialised tasks.[58] Echoing Huntington, in *The Professional Soldier* Morris Janowitz also saw 'skills acquired through intensive training' as the hallmark of the professional army, seeing professional-era officers as similar to other professions such as lawyers and doctors.[59] Interestingly, Huntington specifically argued that the reservist was a caste apart from the new professional military class, claiming that as reservists 'seldom achieve the level of professional skill open to career officers', consequently the reservist 'only temporarily assumes professional responsibility.' Indeed, he went further, positing that the reservist's 'principal function in society lies elsewhere'; an argument that undermines the common perception of reservists' role in building civil-military ties.[60] As a result of this functional difference, Huntington argued that reservists' 'motivations, values and behaviour frequently differ greatly from those of career professionals.'[61]

It is clear that, for Huntington, the origins of the professional military are to be found in expertise, in time spent training, and that because reservists by their very nature do not have the same amount of time as regulars, they are therefore unprofessional. Similarly, Janowitz states bluntly: 'A man is either in the armed forces or not',[62] thereby missing the complex roles and identities of reservists. The views of Huntington and Janowitz are also consistent with Connelly's findings on the British Army's attitudes toward integrating the TA.[63] This definition of professionalism based on status groups with specialised expertise, and, crucially, the amount of time spent undertaking professional activity, is fundamentally at odds with the very concept of reserve service. Indeed, numerous academics have argued that professionalisation – with its shift to a volunteer force encouraging occupational rather than institutional motivations to serve – has caused the demise of the mass-era citizen-soldier, defined by their representativeness of society, their notion of service to the nation, and their primary identity as citizens who are only temporarily in uniform.[64] While these arguments on the death of the citizen-soldier

---

57 Huntington, *The Soldier and the State*, 7, 10.
58 Ibid., 8–9.
59 Janowitz, *The Professional Soldier*, 6.
60 Huntington, *The Soldier and the State*, 17.
61 Ibid., 17.
62 Janowitz, *The Professional Soldier*, xvi.
63 Connelly, V. (2013), *Cultural Differences between the Regular Army and TA as Barriers to Integration*, Unpublished paper prepared for Director, Personnel, MoD. 11.
64 Cohen, E. (2001), 'Twilight of the Citizen-Soldier', *Parameters*, Summer; Moskos, C. (1977) 'From Institution to Occupation: Trends in Military Organization', *Armed Forces & Society*, 4(1), (November); Burk, J. (2007), 'The Changing Moral Contract for Military Service', in

have been challenged,[65] it is clear that within the current attempt to transform the British Army Reserves, there exists an interesting paradox; on the one hand, professional soldiering is still largely defined by full-time service and experience, yet FR20 is seeking to increase the performance of part-time reservists who remain – by the military's own definition – unprofessional citizen-soldiers. As such, this literature provides a rich context to collect data on the juxtaposition between professionalism and the citizen-soldier.

One critique of King's work on the post-Fordist military is that it lacks the wider social and cultural aspects of the post-modernist scholars.[66] Recently, King has convincingly argued that professionalisation 'does not simply involve a change of employment contract between the soldier and the armed forces. It represents a profound transformation of the associative patterns within the armed forces and the solidarities displayed within military units.'[67] He has examined how, at the micro-interaction level, the continued applicability of this skills-based definition of professionalism is evidenced in the successful execution of battle drills and other formalised practices, both individually and collectively.[68] Following Huntington, King argues that competent performance, and the status this generates, defines professionalism in modern militaries. By taking a similar approach to the ongoing attempt to increase the effective performance of the Army Reserve at the sub-unit level, this book investigates the interesting commonalities and contradictions between reserve logistics sub-units and the regular combat forces about which King writes. In relating the literature on the professional military to the British Army Reserve, it adds an additional strand to it. Similarly, by focusing on the logistics sub-unit, this study not only addresses one of the most important areas of FR20, it also complements King's work on the impact of professionalisation in regular combat units.

Cohesion

While the question of why soldiers continue to fight when faced with the horrors of combat has fascinated society since at least the time of Herodotus, it was only in the latter 20th century that social scientists turned their attention to the topic of military group cohesion.[69] Broadly defined, group cohesion has been traditionally

Bacevich, A. (ed.) *The Long War: America's Quest for Security since World War II*, New York: Columbia University Press.

65 Krebs, R. (2009), 'The Citizen-Soldier Tradition in the United States: Has its Demise Been Greatly Exaggerated?' *Armed Forces and Society*, 36(1).

66 Levy, Y. (2010), 'The Essence of the "Market Army"', *Public Administration Review*, 70(3), 379.

67 King, *The Combat Soldier*, 208–209.

68 Ibid., 222–65.

69 Herodotus (2008), *The Histories* (trans. R. Wakefield) Oxford: Oxford Paperbacks.

defined as the 'extent to which members come together to form the group and hold together under stress to maintain the group.'[70] Building on the social-psychological approach of early group interaction theorists such as Charles Cooley and Leon Festinger,[71] Edward Shils and Morris Janowitz's seminal work *Cohesion and Disintegration in the Wehrmacht*, established the classical school of military group cohesion focused on the close interpersonal bonds between small-unit members that motivates them to perform in combat.[72] This view, which has been developed and adjusted to become known as the 'Standard Model', provided the basis for most of the research on military cohesion until the 2000s, when other social psychologists and organisational management scholars began to focus on the motivational influence that commitment to the mission – known as task cohesion – has on military group members.[73] While there is continued debate over which of these components of cohesion is predominant, there is general agreement that cohesion is a multi-dimensional construct whose components can be divided into three distinct categories; a social component, a task component and a group identity component.[74] However, crucially, nearly all of the classical group cohesion studies

---

70 Siebold, G. (2012), 'The Science of Military Cohesion', in Salo, M. and Sinkko, R. (eds) *The Science of Unit Cohesion – Its Characteristics and Impacts*, Tampere: Finnish National Defence University, 44.

71 Cooley, C. (1909) *Social Organization: A Study of the Larger Mind*, New York: Charles Scribner's Sons; Festinger, L., Back, K. and Schachter, S. (1950), *Social Pressures in Informal Groups: A Study of Human Factors in Housing*, New York: Harper.

72 Shils, E. and Janowitz, M. (1948), 'Cohesion and Disintegration in the Wehrmacht in World War II', *Public Opinion Quarterly*, Summer, 286, 291; Stouffer, S., Lumsdaine, A. and Harper, M. et al. (1949), *The American Soldier: Combat and Its Aftermath*, Princeton: Princeton University Press; Henderson, D. (1985), *Cohesion: The Human Element*, Washington, DC: National Defence University Press; Kinzer Stewart, N. (1991), *Mates and Muchacos: Unit Cohesion in the Falklands/Malvinas War*, New York: Brassey's; Wong, L., Koldtiz, T., Millem, R. and Potter, T. (2003), *Why They Fight: Combat Motivation in the Iraq War*, Carlisle Barracks, PA: Strategic Studies Institute, U.S. Army War College; Siebold, G. (2007), 'The Essence of Military Cohesion', *Armed Forces and Society*, 33(2).

73 MacCoun, R. (1993) 'What is known about unit cohesion and military performance', in *Sexual Orientation and U.S. Military Personnel Policy: options and assessment*, Washington, DC: RAND; MacCoun, R., Kier, E. and Belkin, A. (2006), 'Does Social Cohesion Determine Motivation in Combat?', *Armed Forces and Society*, 32(4).

74 MacCoun, R. and Hix, W. (2010), 'Cohesion and performance', in National Defense Institute, *Sexual orientation and U.S. military policy: An update of RAND's 1993 study*. Santa Monica: RAND Corporation, 139–40; Griffith, J. (2007), 'Further Considerations Concerning the Cohesion-Performance Relation in Military Settings', *Armed Forces and Society*, 34(1), 138–39, Beal, D., Cohen R., Burke M. and McLendon, C. (2003), 'Cohesion and Performance in Groups: A Meta-Analytic Clarification of Construct Relations', *Journal of Applied Psychology*, 88(6), 990–91; Dion, K. (2000), 'Group Cohesion from "Field of Forces" to Multi-Dimensional Construct', *Group Dynamics: Theory, Research and Practice*, 4(1); Hogg, M. (1992), *The Social Psychology of Group Cohesiveness: from Attraction to Social Identity*, New York: Harvester Wheatsheaf.

focus on combat forces, and the methods utilised by these schools have mainly been based on interviews or surveys.

More recently, scholars such as King, Ben-Ari, and Hew Strachan, amongst others, have identified other important aspects of cohesion in military units, highlighting the importance of training, communication, and drills.[75] This understanding of cohesion is based not principally on interpersonal bonds, nor motivations, but rather on shared understandings and the practices of military professionalism that enable the group to perform effectively in combat. These authors' emphasis on professionalism, training and collective action is a key addition to the literature, and was arrived at by archival analysis, interviewing soldiers during, and closely observing units in, training and on operations to generate qualitative data. While these different disciplinary and methodological approaches have clearly led to different conceptualisations of cohesion, again even these revisionist cohesion scholars have only focused on regular, combat forces. Indeed, to date there has been no examination of cohesion in logistics units, nor in British reserve units by either the classical or revisionist cohesion scholars.

Logistics Sub-units

Due to its close relationship to the army's operational capability the transformed reserve is designed to deliver, the emphasis on deploying Army Reserve sub-units is worthy of further discussion here. In the past, although some infantry and medical company groups have collectively deployed on operations, this has been the exception rather than the norm. The pattern of mobilisation of the TA, in both the infantry and its supporting services, has usually been one of the 'intelligent mobilisation' of individuals who volunteer to serve on operations by backfilling regular units, rather than deploying fully formed reserve units.[76] Thus, by deploying formed units and sub-units, FR20 aimed to significantly change how the TA was used on operations. This requirement to deploy formed sub-units presents new challenges for the Army Reserve in terms of delivering the capability and readiness expected of it under FR20. Indeed, given FR20's emphasis, the sub-unit provides a particularly useful level of analysis for investigating FR20 as it is at this level that the transformation has been focused.

---

75 Strachan, H. (2006), 'Training, Morale and Modern War', *Journal of Contemporary History*, 41(2); Ben-Shalom, U., Lehrer, Z. and Ben-Ari, E. (2005), 'Cohesion During Military Operations: A field study on combat units in the Al-Aqsa Intifada', *Armed Forces and Society*, 32(1); King, A. (2006b), 'The Word of Command: communication and cohesion in the military', *Armed Forces and Society*, 32(1). King, *The Transformation of Europe's Armed Forces*; King, *The Combat Soldier*, 208–338.

76 Dandeker, C., Greenberg, N. and Orme, G. (2011) 'The UK's Reserve Forces: Retrospect and Prospect', *Armed Forces and Society*, 37(2).

FR20 placed heavy emphasis on the outsourcing of logistics capabilities to the reserve. Understanding how logistics organisation and practice has transformed in the past 15 years is therefore central to understanding the rationale behind FR20. Indeed, the transformation of the reserves and in particular its logistics component cannot be understood without recognising the drastic changes in how logistics is now delivered. FR20's central focus on the reserve logistics component is critical to my arguments as it provides the rationale for examining reserve logistics sub-units. Army2020 drastically reduced the size of the regular army's logistics component to save costs. For example, the regular army's Royal Logistics Corps (RLC) lost two regiments, while the Royal Electrical and Mechanical Engineers (REME) lost one battalion as result of the policy.[77] However, as detailed above, FR20 outlined significant changes to the capabilities expected of reserve logistics sub-units, stating 'Greater reliance will be placed on the Reserves to provide routine capability … primarily in … combat service support (such as logistics …)'.[78] A central organising principle of FR20 is therefore the outsourcing of logistics capability previously held in regular forces to the reserves to save costs. This cannot be overstated: the main organisational focus of the FR20 transformation was on Combat Service Support (CSS, or logistics). Meanwhile, other reserve units have been formed to deliver bespoke logistic capabilities. A central tenet of this increasing reliance on the reserve component is that CSS reserve organisations at the sub-unit (company or squadron) level will be held at a higher level of readiness and must be capable of operating with their regular counterparts. In short, reserve logistics sub-units are to deliver more of the capability previously provided by regular units. Thus, FR20 also represents an attempt to transform the role, capability and deployability of logistics sub-units.

Indeed, communication with Major General Kevin Abraham, Director General Army Transformation (the officer responsible for implementing FR20), indicated that one of the greatest risks to successfully implementing FR20 lay with the logistics component, and within this group, the REME and RLC; the regular army will be more dependent on the ability of these elements to deploy quickly and perform effectively.[79] The REME provide the army with its mechanics, and are specialists in heavy vehicle and helicopter recovery, repair and maintenance. The RLC has a myriad of responsibilities, from the transportation of supplies by road, sea and air, to the provisioning of food, operating ports and delivering post, amongst many others. The ability of REME and RLC sub-units to meet the new readiness requirements is therefore central to FR20's aims and important to its success, which

---

77   BBC News (2012), 'Army to lose 17 units amid job cuts', 5 July, available at https://www.bbc.co.uk/news/uk-18716101

78   *Future Reserves 2020*, 22.

79   Interview, Director General Army Transformation, Major General Kevin Abraham, Andover, 14 January 2014.

in turn underpins Army2020. By contrast, the reserve combat component was not deemed to present such a major organisational challenge to FR20.[80]

The British Army Reserve

Not only are these complementary and interrelated perspectives highly useful in assessing FR20's effectiveness, they also address gaps in the literature on reserve forces in general and on the British reserves in particular. Reflective of their increased importance to post-Cold War security, numerous studies have emerged that consider these forces from national perspectives (the US, France, Australia and Israel in particular).[81] While a central theme running through some of these works is the impact that the professionalisation of regular armies has had on respective reserve forces, none of these studies conducted in-depth examinations of how periods of organisational change have shaped this professionalisation, nor their effects on cohesion or conceptualisations of professionalisation in the reserves. Instead, they are organisational summaries of the current state of play within their respective forces and do not provide a deep, sociologically-informed analysis of the forces in question. Another strand of the wider reserve literature focuses on identity, motivation and other specific issues.[82] These also do not address the military transformation, professionalism, or cohesion literature.

---

80 Interview, Director General Army Transformation, Major General Kevin Abraham, Andover, 14 January 2014.
81 Ben-Ari, E. and Lomksy-Feder, E. (2011), 'Epilogue: Theoretical and Comparative Notes on Reserve Forces', *Armed Forces and Society*, 37(2); Perliger, A. (2011), 'The Changing Nature of the Israeli Reserve: Present Crises and Future Challenges', *Armed Forces and Society*, 37(2); Smith, H. and Jans, N. (2011), 'Use Them or Lose Them? Australia's Defence Force Reserves', *Armed Forces and Society*, 37(2); Weber, C. (2011), 'The French Military Reserve: Real or Abstract Force?', *Armed Forces and Society*, 37(2); Griffith, J. (2011), 'Contradictory and Complementary Identities of US Army Reservists: A Historical Perspective', *Armed Forces and Society*, 37(2); Griffith, J. (2009), 'After 9/11 What Kind of Reserve Soldier?', *Armed Forces and Society*, 35(2); Griffith, J. (2011), 'Decades of transition for the US reserves: Changing demands on reserve identity and mental well-being', *International Review of Psychiatry*, 23(2).
82 Lomsky-Feder, E., Gazit, N. and Ben-Ari, E. (2008), 'Reserve Soldiers as Transmigrants: Moving between the Civilian and Military Worlds', *Armed Forces and Society*, 34(4); Griffith, J. (2009), 'Being a Reserve Soldier: A Matter of Social Identity', *Armed Forces and Society*, 36(1); Vest, B. (2013), 'Citizen, Soldier, or Citizen-Soldier? Negotiating Identity in the US National Guard', *Armed Forces and Society*, 39(4); Ben-Dor, G., Pedazhur, A., Canetti-Nisim, D., et al. (2008), 'I versus We: Collective and Individual Factors of Reserve Service Motivation during War and Peace', *Armed Forces and Society*, 34(4); Griffith, J. (2008), 'Institutional Motives for Serving in the US Army National Guard', *Armed Forces and Society*, 34(2); Griffith, J. (2012), 'Correlates of Suicide Among Army National Guard Soldiers', *Military Psychology*, 24(6).

The same is true of sustained academic study of Britain's reserve forces, and of the TA in particular. In 1975 Hugh Cunningham, and later Ian Beckett in 1982, both traced the origins and evolution of the Volunteer movement that preceded the TA, while Peter Dennis' useful study of the TA between 1906–1940 was published in 1987.[83] More recently, K.W. Mitchinson has examined the Territorials' formative years, its role in the Great War, and on the home front.[84] More pertinent to this study, in 1990 Walker published the most recent and in-depth organisational analysis of the TA.[85] He noted that the part-time, voluntary nature of service in the TA meant that it was fundamentally distinct from the regulars, and identified a number of organisational paradoxes. Tellingly, this included the contradiction between a militarily-capable force generated by intensive training and the lack of time the organisation – due to its part-time nature – had to achieve this. Thus, as well as the many cultural and organisational differences between the TA and the regulars, Walker recorded high variations in culture and capability across TA sub-units and a 30 per cent turnover in personnel per year.[86] This created another contradiction whereby the need to continually train new personnel resulted in major retention issues. As a result, Walker argued that the force was predominantly a reactive force of last resort, never really ready for war, and would likely have been unable to fulfil its Cold War mission of providing support to first-line NATO troops in Western Europe. Perhaps most significantly, he noted that the voluntary and part-time nature was conducive to organisational stasis; as one TA officer remarked: 'There is nothing in the TA you can effect immediately. It takes 5–10 years [to make changes].'[87]

However, written at the end of Cold War, Walker's insightful study is now over 30 years out of date. Furthermore, while Walker's analysis is interesting and comprehensive, it lacks a theoretical framework beyond brief reference to organisational theory literature.[88] Consequently, his analysis could be argued to lack conceptual depth and it does not attempt to inform a wider analysis of reserve forces in general; the series of organisational 'paradoxes' Walker outlines in the TA are therefore left to the reader to decide if these are unique to the TA. In short, Walker's study tells us how the TA was in 1990, without placing it within the broader sociological literature on professionalism, cohesion, and transformations.

---

83 Cunningham, H. (1975), *The Volunteer Force: A Social and Political History, 1859–1908*, London: Croom Helm; Beckett, I. (1982), *Riflemen Form: A Study of the Rifleman Volunteer Movement, 1859–1908*, Aldershot: The Ogilby Trust.

84 Mitchinson, K. (2008), *England's Last Hope*: *The Territorial Force 1908–14*, London: Palgrave; Mitchinson, K. (2005), *Defending Albion: Britain's Home Army 1908–1919*; London: Palgrave; Mitchinson, K. (2014), *The Territorial Force at War 1914–16*, London: Palgrave.

85 Walker, W. (1990) *Reserve Forces and The British Territorial Army*, London: Tri-Services.

86 Ibid., 3, 65.

87 Walker, *Reserve Forces*, 65.

88 Ibid., 10–11.

It is therefore more of a snapshot of the TA in time, rather than a theoretically-driven piece of analysis. And the major question remains: how has the TA changed since then? More recently, Beckett published a historical overview of the TA in 2008, charting the changing roles and organisation of the Territorials over the first 100 years of their service. While illuminating, this is historical rather than sociological analysis and it does not include the most recent attempt to transform the Army Reserves.[89] In summary therefore, the major studies of the TA are out of date and do not address recent wider sociological debates about professionalism, cohesion and transformation in modern Western reserve forces, nor indeed, their logistics components.

Even the most recent research on the TA/Army Reserve does not specifically address these questions. While Christopher Dandeker et al.'s 2011 article does outline the bureaucratic debates around the changing role of the British reserves in general, it does not focus on the Army Reserve in particular and, as it predates the FR20 transformation, it lacks detail as to its exact nature and the possible implications for the Army Reserve. Organisational culture academics have also studied the traditional clash of cultures between the regulars and the Territorials, most recently at the MoD's behest. Charles Kirke's small-scale study examined the Regular Army's perception of closer integration of the TA in 2008, and found that significant issues and cultural differences need to be overcome before any such policy could be effective.[90] Vince Connelly's interesting 2013 study of perceptions amongst regular and reserve soldiers, also undertaken for the MoD, drew similar conclusions, outlining the many cultural and practical barriers to integration that are undermining efforts to impose the FR20 transformation on both components of the army.[91] While Tim Edmunds et al. have recently published an overview of FR20 and some of the issues identified to date, especially in terms of recruitment, this paper lacks a recognition of the intensely political origins of FR20 and does not attempt to assess it in terms of a transformation.[92] Very recently, two senior reservists, Jeremy Mooney and John Crackett, published a brief overview of the AR's recent history, without any reference to theory.[93] Decisively, no study has been primarily concerned with the military capability aspects of FR20, and especially this project's focus on capability, transformation, professionalism, and cohesion. The sub-unit has also been ignored. Overall, therefore, this book is well-

---

89   Beckett, I. (2008), *Territorials: A Century of Service*, Plymouth: DRA.
90   Kirke, C. (2008), 'Issues in integrating Territorial Army Soldiers into Regular British Units for Operations: A Regular View', *Defense and Security Analysis*, 24(2).
91   Connelly, *Cultural Differences*.
92   Edmunds, T., Dawes, A., Higate, P., Jenkins, N. and Woodward, R. (2016), 'Reserve forces and the transformation of British military organisation: soldiers, citizens and society', *Defence Studies*, 16(2).
93   Mooney, J. and Crackett, J. (2018) 'A Certain Reserve: Strategic Thinking and Britain's Army Reserve', *The RUSI Journal*, 163(4).

sited to update the literature on the British reserves in general by contributing new data on FR20, while also using this evidence to contribute to wider, related debates on transformations and reserve professionalism and cohesion.

FR20's attempt to transform the Army Reserve's logistics raises a number of interesting questions about how the policy originated and how these origins determined its implementation. In order to understand the context of FR20 and hence its likely impact, the book utilises a number of complementary but different literatures and methods to understand different aspects of the FR20 'problem'. To address these interrelated issues, this book takes a post-Fordist approach and applies it across three levels of analysis. At the macro-level, post-Fordism provides a theoretical framework for both the transformation of Western militaries in general and their military logistics in particular. As I have shown, this is critically important in understanding the organising principles behind FR20. In making this argument in Chapter 3, I conceptually advance the literature on military logistics by detailing how logistics have been transformed in the past 15 years around post-Fordist principles. At the meso-level, the post-Fordist literature fuses with that on military transformation and professionalism, which run throughout this work. At the micro-level, these fuse with the literature on cohesion, informing Chapter 6 especially. These distinct but related literatures provide useful and complementary tools for examining the transformation of the British Army Reserve both qualitatively and quantitatively across three levels of analysis.

The Research[94]

My evidence is drawn from an ESRC-funded, Army-sponsored research project conducted between 2012 and 2016 that examined FR20's origins, evolution and impact. It adopted an interdisciplinary, mixed-methods approach to address different aspects of the FR20 'problem', using primary and secondary sources, elite individual and group interviews, longitudinal surveys, and fieldwork observations. Primary and secondary sources were used to address the questions concerning FR20's context. Once this had been completed, qualitative data on reserve logistician's perceptions of the impact of the transformation on their sub-unit's effectiveness and cohesion was collected through individual and group interviews and field observations. This was simultaneously complemented by quantitative data collection involving three surveys of logistics reservists. These methods were chosen to generate the richest qualitative and quantitative data to ensure triangulation and increase the validity of the evidence. Data was further triangulated with official statements, analysis of government documents, autobiographies, newspaper archives and academic texts.

---

94 This is a shortened description of method and survey design, please contact the author for further details.

Individual interviews were conducted to gain insight into how the FR20 policy was formed, the exact intent of the plan in relation to reserve logistics units, and to give context to the data collected in the group interviews. Informal interviews also took place during the fieldwork. Overall 16 formal and nine informal interviews were conducted. These 25 interviews consisted of a serving minister; two former defence ministers; a serving Chief of the Defence Staff, a former Chief of the General Staff; three generals involved in different aspects of FR20; and two brigadiers and two colonels involved in logistics and the reserve component. Others included sub-unit commanders, other officers, non-commissioned officers (NCOs) and private soldiers. These were selected on the case relevance basis; people whose positions and roles indicated they would be informed on the subject. Furthermore, a total of 14 group interviews with over 150 reservists of all ranks were conducted in four logistics sub-units selected to represent a broad spectrum of locations and experiences of FR20. Of note is that the group interview sample was intentionally weighted toward sub-units that had undergone organisational transformation. To improve reliability, the group interview data was transcribed and then analysed to identify common response themes using the NVivo coding software. Fieldwork was also conducted to collect qualitative data on the importance of collective training in reserve logistics units, and to triangulate this with the interview and quantitative data. The Ministry of Defence (MoD) Research Ethics Committee approved the project and all participants consented to taking part. Most participants were anonymised, including all group interviews, but those senior officials quoted gave their consent.

The cohesion and readiness survey results presented in Chapter 6 were based on Siebold and Kelly's Platoon Cohesion Index, and also utilised some items from Reuven Gal's and James Griffith's morale and readiness surveys.[95] All questionnaires have been used to assess these factors in a number of militaries. A number of additional items covering attitudes toward FR20 were added. Approximately 1,500 personnel from a total 43 units were approached to participate, and there were 427 valid responses. The 2015 response rate was 29 per cent, which is consistent with Army Reserve responses in the 2018 Tri-Service Reserve Continuous Attitudes Survey (30 per cent). The 2016 response rate was too low for statistical significance and forced the use of only three selected sub-units with strong response rates and internal validity. On average RLC personnel represented 72 per cent of responses, with REME 28 per cent. This is representative of the RLC and REME reserve population, and, with missing data excluded testwise, a chi square test for goodness-of-fit confirmed this (1, n = 427) =.39, p =.53.

---

95   Siebold, G. and Kelly, D. (1988) *The Development of the Platoon Cohesion Index*, Washington, DC: Army Research Institute; Griffith, J. (1988) 'Measurement of Group Cohesion in U.S. Army Units', *Basic and Applied Social Psychology*, 9(2); Gal, R. (1986), 'Unit Morale: From a theoretical puzzle to an empirical illustration – an Israeli example', *Journal of Applied Social Psychology*, 16(6).

The research focus on the reserve logistics component was suggested by the army as this area of the FR20 transformation was deemed at higher risk than in the combat arms. This was for a number of reasons, but mainly because the changes that some logistics units had to undergo in order to provide the required capability – including forming new units, changing base locations or re-roling into a new trade – represented major organisational changes that would require considerable time and effort to implement. Even taking this into account, the transformation of the army reserve logistics component was not viewed as guaranteed. As a result, it must be made clear from the outset that this book used a sample of logistics units of which some had undergone profound organisational change as a result of FR20. This sampling was intentional as the research was firmly based in both the logistics component and the transformation literature and sought to understand factors affecting change in these units. It is therefore recognised that the logistics reserve component may have proved more difficult to transform than other components, where experiences of transformation may have been more positive. For example, it was beyond the scope of this study to examine in detail the experiences of infantry units. Nevertheless, I have incorporated data from regular and reserve infantry and logistics units for comparison. Moreover, while any assessment of a six-year policy before it has fully completed has its weaknesses, and further developments no doubt will occur, I have made every effort to identify enduring issues that are likely to shape FR20's impact in the future.

Overview

This book examines the origins, evolution and impact of FR20 as an attempt to organisationally transform the British Army Reserve, using its logistics forces as an evidential base. This attempt to transform the Army Reserve's logistics raises a number of interesting questions about how the policy originated and how these origins determined its implementation. This study primarily takes a post-Fordist approach to examine FR20, while also utilising the literature on military logistics, professionalism and cohesion. Chapter 2 assesses the strategic, political and organisational factors that have influenced previous attempts at reserve reform. While these previous reforms are centrally important to understanding the historical context for the current transformation, they allow me to highlight that while reserve transformations are often driven by similar political, financial and strategic factors, the organisational solutions that FR20 has adopted are unique. It also shows that organisational resistance curtailed most previous attempts at reserve reform. To provide the conceptual framework, Chapter 3 then utilises a post-Fordist framework to conceptualise the processes through which this drastic change has occurred. Post-Fordist processes are centrally important to understanding FR20's design and aims, but also to understanding its impact on the reserve logistics units who must now provide much of the capability outsourced from the regulars. Crucially, the organisation of forces around post-Fordist principles also provides a historically novel solution to the recurrent organisational problems experienced in past periods of reserve reform. The chapter also originally contributes to the military logistics literature by detailing the post-Fordist principles around which modern Western military logistics structures and practices have been designed, and how, ultimately, these principles have shaped wider force structures.

Building on this evidence, Chapter 4 asks how and why did FR20 come to be implemented? It argues that FR20's origins are best understood by examining the intensely intra-party political and ideological dynamics inside the governing Conservative Party. These origins combined with, and exacerbated, regular-reserve intra-service rivalry to shape the development and implementation of an ad hoc and politically opportunistic defence policy not grounded in the organisational realities of the Army Reserve. As such, from the outset the FR20 transformation was an improbable mission. Once this context has been discussed, this book investigates FR20's impact in the area where it – by its own definition – matters the most: in reserve logistics sub-units. Chapter 5 examines the ability of sub-units to meet the increased 'hard' operational capability requirements demanded by FR20's post-Fordist approach. It highlights the bottom-up organisational factors impeding the transformation, while also assessing areas of FR20's success to date, namely opportunity, training course availability and integration. However, crucially, and controversially, it shows how the political origins of FR20 have resulted in an overemphasis on, and the politicisation of, recruitment.

Aside from capability and perceptions of cohesion and readiness, FR20 also marks a potentially decisive change in the relationship between the regular army and the reserves; between a full-time professional army and what has traditionally been a part-time force of last resort. In short, FR20 is attempting to professionalise the reserves. These attempts to professionalise citizen-soldiers, and especially those in logistics trades, provides a new evidential base to investigate the impact of FR20 on cohesion, and more broadly, to compare them with the nature of cohesion in modern, post-Fordist regular combat forces. Chapter 6 firstly presents statistical evidence of reservists' perceptions of cohesion, morale and readiness within the context of FR20, and longitudinally examines changes in these perceptions as FR20 progresses. It shows that while perceptions of cohesion and morale are relatively high, they have not increased as a result of FR20, and confidence in FR20 succeeding is decreasing over time. The second part of the chapter discusses the softer, cultural impacts of FR20 on the selected units, allowing arguments about the nature of cohesion in logistics and reserve forces in general to be made. Building on these arguments, in the concluding chapter I draw together the study's findings, before widening the scope to discuss what the experience of FR20 tells us about recent British civil–military relations, and the major changes in British society since the 1960s.

My central argument is that the intra-party political origins of FR20, intra-service rivalry, and the Army Reserves' organisational nature, have undermined the policy's ability to deliver the key military capabilities it envisaged of the reserves. However, in some important cultural/normative aspects, FR20 is slowly transforming them. Thus, my overall argument is that FR20 has been, and will be, a 'partial transformation'. It is struggling to deliver its central aim of increasing reserve logistics capability in the timeline required, but conversely, it is gradually changing cultural elements of the Army Reserve. Some micro-level associative patterns, though, are likely to prove more resistant to change.

In making this argument I make three major contributions to defence debates. Firstly, I provide new and detailed empirical analysis of the organisational origins and evolution of FR20. This analysis highlights incoherent defence planning in the Cameron era. Secondly, in order to understand FR20's heavy focus on increasing reserve logistics capability and its impact on logistics units, I detail how Western military logistics structures and practices have recently transformed in line with what have been termed 'post-Fordist' principles. In doing so, I challenge much of the traditional military logistics literature. Finally, by examining the impact of FR20 in sub-units, I contribute new evidence on the nature of cohesion in reserve forces, while highlighting how the distinct nature of reserve service has resulted in a partial transformation. The conclusion discusses how this partial transformation has been reconciled with FR20's original aims, and what this transformative attempt tells us about modern British civil–military relations and British society in general. I argue that although FR20 has been a 'partial transformation' due its intra-party political origins, intra-service rivalry and organisational frictions, some of its failings as a transformative attempt are due to its failure to understand how British society itself has transformed.

Chapter 2

Balancing Budgets, Strategy and Recruitment: Previous Reserve Transformations

As outlined in the last chapter, the British Army, and the Army Reserve, is currently in a period of profound organisational transformation. On the one hand, this attempt to re-organise the Army Reserve for complex and diverse 21st century missions appears to be a response to the strategic uncertainty of an increasingly globalised world.[1] As is discussed in the following chapter, the desire for more professional and adaptable reserve forces is symptomatic of the post-Fordist approach to military organisation that has developed in this era, primarily in order to generate greater efficiencies. Understanding this approach is central to an understanding of the attempted transformation of reserve logistics capability. This chapter, on the other hand, examines the historical context of previous attempts to change British reserve land forces to show how the current attempt at transformation is perhaps less novel than it initially appears. It shows that the impetus for transforming the army's reserve forces has remained remarkably similar during different periods. Indeed, the interplay between budgetary constraints, strategic rationale, and recruiting the reserve force has heavily influenced the decision to implement – and ultimately the effectiveness of – the previous three major attempts to transform the reserves. While this chapter shows that attempts to reform the reserve are cyclical – they represent an attempt to respond to changed strategic, economic and operational circumstances – crucially, it also shows that the organisational solutions to the problems these changed circumstances present are historically specific. Moreover, each previous attempt at reserve transformation has been politically-imposed rather than undertaken by the army of its own volition.

How stakeholders influence policy outcomes has been excellently highlighted by Graham Allison in his dissection of American and Soviet strategic rationale, bureaucratic politics, and organisational process during the Cuban Missile Crisis,

---

1 HM Government (2010) *Securing Britain in an Age of Uncertainty: The Strategic Defence and Security Review*, 9, paragraph 1.4.

on which Posen heavily drew in his seminal study on transformation.[2] Both Allison and Herbert Kaufmann, amongst others, have argued that organisations transform when intense pressures build up around them to force change in order to ensure their continued viability.[3] Transformational leaders and the acquiescence of internal elites are needed to drive through this change, but in meeting organisational resistance, these changes are negotiated and modified. The net result is that the organisation adapts. But these adaptations are not dramatically new, they are instead a re-booted version of procedures the organisation is already familiar with. Organisational change thus occurs through, and is modified by, the interplay of its parts. This chapter seeks to identify the key roles of stakeholders in previous reserve army reforms and explain how they shaped their outcomes, in order to better understand the current transformation. This historical analysis is central to my argument as not only does it show the continuities with the past, it also highlights how the latest, post-Fordist approach is an organisationally novel attempt to solve enduring fiscal, strategic and organisational problems.

The Cardwell-Childers Reforms: A First Attempt at Integration

Although the origins of the Army Reserve date back to the fyrds of the Anglo-Saxon period, it was not until Henry II's Assize of Arms in 1181 and Edward I's Statute of Westminster in 1285 that the military obligation of freemen to defend the community became enshrined in England. The need for continental armies in the mid-16th century saw the first of the Militia statutes in 1558, effectively incorporating the Militia into a formal existence, which remained in place despite periods of repeal until the twentieth century.[4] Growing government regulation and the gradual professionalisation of the army and its reserve forces over this period has been charted by numerous scholars.[5] By the mid-1860s, Britain's army reserve forces were essentially organised in two systems whereby those who volunteered for the Militia would complete a few months' initial training and then return to civilian life on the understanding that they would undertake a few weeks' annual refresher training. The Militia's mission was defence against invasion (although it was used in a public order role when required) and those serving in it, who were mainly drawn from agricultural areas, signed up for five years, on the condition that they could not be deployed overseas nor posted outside their regiment. The

---

2   Allison, G. (1971) *The Essence of Decision: Explaining the Cuban Missile Crisis*, Boston: Little and Brown.

3   Kaufmann, H. (1971) *The Limits of Organisational Change*, Tuscaloosa: University of Alabama Press.

4   Beckett, *Territorials*, 3–4.

5   Mallinson, A. (2011) *The Making of the British Army*, London: Bantam; Beckett, *Riflemen Form*; Cunningham, *The Volunteer Force*.

Yeomanry was the cavalry arm of the Militia and served under the same conditions. The other part of the system was represented in the more urbane Volunteers who had been created in 1859–60 as a result of the public's largely imaginary fears of an imminent French invasion. The Volunteers consisted of mainly riflemen, gunners and engineers who, like the Militia and Yeomanry, had been recruited locally and could also not be compelled to serve overseas. However, unlike the Militia who were more closely controlled by the state, the Volunteers had originally funded and equipped themselves, representing 'the military expression of the spirit of self-help, Victorian capitalism in arms.'[6]

Although on paper the Militia was 130,000 strong and the Volunteers numbered over 160,000 by 1868, both the Crimean War and the Indian Mutiny had exposed the inability of Britain's reserve system to mobilise the required numbers of second line troops when the army deployed overseas.[7] Moreover, by the late 1860s, the Yeomanry and especially the Volunteers were in decline, as both forces fell out of fashion with the landed gentry and upper middle classes that supplied them with the officers and recruits and, decisively, the donations they required to function.[8] As the problem worsened, prominent Volunteers in Parliament began lobbying for increased government funding to make up the budget shortfalls of the supposedly self-sufficient Volunteers.

While the Volunteers had cost the state just £3,000 (£250,000 today) in 1860, by 1897 total government expenditure on them was £697,000 (over £70 million), even though the size of the force remained relatively stable during this period. Moreover, this growing drain on the government's coffers was occurring precisely at a time when the army was being increasingly criticised, both by the Radicals in Parliament who detested the aristocratic nature of the army – especially the practice of purchasing commissions – and by some officers concerned about the army's and the reserve's military effectiveness.[9] Such criticism was justified: between 1864 and 1869 spending on the Army and Ordnance far exceeded that for any other branch of government, eclipsing funding for the Navy by an average of £4 million per annum (£420 million today).[10] A considerable proportion of these costs paid for soldiers' pensions rather than for effective military capability, while doubts about the Volunteer's effectiveness had already led to an attempt in 1867 to create an 'Army Reserve' of 20,000 men that had failed miserably.[11] As such, there was a growing realisation within government that the army and the reserves needed to be reorganised to guarantee both better efficiency and value for money. It is

---

6   Cunningham, *The Volunteer Force*, 1.
7   French, D. (2005) *Military Identities: The Regimental System, the British Army and the British People*, Oxford: Oxford University Press, 12.
8   Walker, *Reserve Forces*, 13.
9   Beckett, *Riflemen Form*, 138.
10 French, *Military Identities*, 12.
11 Ibid.

important to stress that this political desire to reduce the military budget was the primary driver of the forthcoming reforms. But the drive to make these forces more economical and efficient also fused with the strategic situation, coming as it did after a major war and during continued colonial withdrawal from Canada, Australia and New Zealand.[12] Indeed, one of the most immediate impacts of the subsequent reforms was the reduction in colonial garrisons by over 25,000 men between 1869 and 1871, at considerable savings to the War Office.[13]

Yet the withdrawal from the New World was not the only strategic rationale to influence the forthcoming organisational reform of the army and reserves. Poorly trained Union volunteer units had not performed well in the opening stage of the American Civil War. This was in stark contrast to Prussian militia units whose recent victories in Europe had demonstrated the importance of thorough reserve training and discipline. Moreover, as British studies had concluded, the Prussian system of localised recruitment and the pairing of Line, Reserve and Landwehr (militia) units allowed fast mobilisation and rapid expansion of the Prussian army in wartime. Indeed, such was the strategic and organisational success of this system in defeating the Austrian and French armies that, echoing Farrell's observations on military emulation, it formed the blueprint for the reforms instigated by the Secretary of State for War, Edward Cardwell. General Garnett Wolseley was instrumental in driving the following wider reform of the army and ensuring the reform of the reserves was integrated with it.[14]

However, the Cardwell reforms were also as much a product of the dominant ideology at the time as they were the economic and strategic context. The 1867 Reform Act had committed the Liberal government to a wider economic and social programme by extending the franchise, and the reform of the army became a means by which the Liberals hoped the working class would be lifted from poverty into respectability and thus become better integrated into the political life of the nation. As David French has shown, the Radicals hoped the army would become 'a powerful instrument for national education in a large and high sense' as a result of the introduction of new short-service contracts.[15] This ideological element of the reforms was taken up by Gladstone himself, who argued that the introduction of new local depots would 'diminish to a minimum immorality in the standing army', while some army officers also argued that army service would help create a 'more perfect man and a better citizen.'[16] Political ideology was therefore clearly mobilised to support reserve reform.

---

12 Spiers, E. (1994) 'The Late Victorian Army 1868–1914', in Chandler, D. and Beckett, I. (eds) *The Oxford History of the British Army*, Oxford: Oxford University Press, 191.

13 French, *Military Identities*, 14.

14 Brazier, J. (2012) '"All Sir Garnet!" Lord Wolseley and The British Army in the First World War', *Military History Monthly*, May.

15 French, *Military Identities*, 26.

16 Ibid., 26–32.

It was within the context of these economic, strategic and ideological debates that the reform of the reserves was shaped. The Cardwell reforms – which began in 1868 with attempts to abolish the purchase of commissions – are best known for enshrining the regimental system into the army's organisational structure through the implementation of the policies of localisation and pairing with reserve units. However, the reforms were wide in scope, with the 1870 Army Enlistment Act reducing the period of service from 21 years to twelve, with most men passing into the Army Reserve after six years of service. This act was vital, as not only did it cut the pension bill, it created a system, on paper at least, by which the army could be expanded rapidly in times of crisis, while it was also hoped that shortening the terms of service would attract a better quality of recruit. Meanwhile, the Localisation Act of July 1872 and General Order 32 of 1873 created linked and localised line infantry regiments. Linking saw two-battalion infantry regiments become the norm, with one serving overseas and the other at home, while localisation divided the country into 66 sub-districts, each with its own pair of linked battalions and own permanent depot. This depot was to be shared with at least two local Militia battalions and any already existing local Volunteer battalions. The rationale behind these reforms was clear; recent evidence from the Prussian and Confederate armies suggested that locally-recruited battalions had better morale and discipline,[17] while in co-locating the headquarters of both regular and reserve units, Cardwell hoped that training in close proximity with regulars would increase the efficiency of the reserves while also encouraging them to join their full-time counterparts.[18] Similarly, Wolseley's visits to Confederate troops fighting in the American Civil War had convinced him of the need for more musketry and staff training.[19]

The drive for better integration of the reserves was complemented by other measures to increase state control over them and their efficiency. Lord Lieutenants' jurisdiction over the Volunteers was replaced by the Secretary of State for War's, and a proficiency certificate for Volunteer officers and NCOs was introduced. Volunteer adjutants were to be phased out and replaced by their regular counterparts and permanent staff instructors to boost capability.[20] This nascent attempt to professionalise the Volunteers was also evident in the introduction of a musketry bonus for soldiers who met the increased standard, and in the amount of mandatory unit training required before Volunteer units could receive their increasingly important capitation grants from the government, upon which they relied for survival.

While these reforms were well-intentioned, predictably they faced intense criticism from numerous stakeholder groups opposed to the changes. As French

---

17 French, *Military Identities*, 15.
18 Biddulph, R. (1904) *Lord Cardwell at the War Office: A History of His Administration 1868– 1874*, London: John Murray, 173.
19 Brazier, "'All Sir Garnet!'", 37.
20 Beckett, *Riflemen Form*, 129–30; French, *Military Identities*, 15.

has shown, in terms of the wider army, the switch to short-service created an unforeseen recruitment and retention problem, with the number of soldiers needed per annum as a result of the introduction of short-service doubling by 1879.[21] This was complemented by a steep rise in deserters, the reduction in the quality of NCOs – who now had less experience – and an overage officer corps.[22] Meanwhile, the Volunteers mobilised to resist the steady incursion of the regulars into their domain. Volunteer Adjutants rejected the introduction of the proficiency cert as demeaning, and also clashed with the government over the pegging of the capitation grant to the two-thirds unit turnout required at parades. Most vociferously, they attempted to reject their replacement by regulars, becoming 'something of a pressure group in parliament'.[23]

Added to the recruiting problem and internal dissent, strategic imperatives heaped organisational pressures on the army and Cardwell's plan for the reserves. The Ashanti, Zulu, Afghan, First Boer and Egyptian campaigns tested the linked battalion system to the limit, with home battalions essentially becoming feeder units for their sister battalions fighting abroad. As a result of overstretch, the cohesion-destroying practice of cross-posting soldiers between regiments – precisely what Cardwell's linking had been designed to end – became common again. Meanwhile, it was also apparent that the transfer of former soldiers into the new Army Reserve could not match the need for soldiers to serve abroad because many of these campaigns could not be classed as the 'grave national emergency' required to mobilise them. Thus, by the late 1870s it was becoming clear that while Cardwell's reforms had changed the army and reserves on paper, in reality it had not been as thorough a reformation as intended.[24] While localisation had been achieved, it had been at the expense of organisational balance and recruitment. These issues would need to be addressed by the next war minister, Hugh Childers.

The fact that Childers needed to undertake any reforms at all offers stark proof of the failure of Cardwell's plans. Indeed, by 1881 criticism of Cardwell's efforts had become so vociferous that a report by a committee of general officers recommended abandoning the system of linking entirely and replacing localisation – which hinged on a commitment to only post a soldier within his regiment – with a 'general service' contract. However, crucially, most of the £3.5 million (about £310 million today) allocated by Cardwell to build the regimental depots had now been spent, and there was no way Childers could abandon such a costly programme. Faced with this economic and political reality, he continued it, pushing localisation further by amalgamating the linked battalions into new territorial regiments now named after the locality they recruited from. Militias made up these regiments' third and fourth battalions, with Volunteer units also taking the new regiments' territorial

---

21 French, *Military Identities*, 16.
22 Ibid., 18–19.
23 Beckett, *Riflemen Form*, 131.
24 Spiers, 'The Late Victorian Army', 191–92.

names. To address the recruitment and retention problem, Childers lengthened the terms of service to seven years and reduced reserve liability to five, while also improving soldiers pay, promotion terms, and pensions. He also brought the Draconian discipline system more in line with Liberal principles.[25] The continued drive for efficiency and professionalism also affected the Volunteers, with further mandatory requirements for battalion drills in camp, the introduction of a voluntary exam for field officers and the introduction of uniform regulations, the latter of which especially prompted much agitation from officers over the expense incurred and ignorance of history the new uniforms represented.[26]

However, while Childers largely completed the process of army reform begun by Cardwell, his Militia and Volunteer reforms were less successful. Intensified linking and territorialisation did not improve relations between the regulars and their reserve counterparts in all units, mainly due to ongoing mutual professional suspicion; militiamen training at the Suffolk Regiment's depot before the Boer War were derided as 'half-soldiers' by their regular counterparts.[27] There was also strong Militia agitation against moving their headquarters to the new regimental depots as they believed their headquarters were already local enough to sustain recruitment and identity.[28] The fact that Militia units also lost the ability to train their own recruits at the depots was another source of conflict. Moreover, class-related social divisions between the army and the Volunteers meant that the latter did not provide the steady flow of recruits into the army in the numbers that Cardwell and Childers had hoped localisation, territorialisation and subsequent integration would prompt. However, as French has shown, the Militia did join the army in significant numbers – about one third of recruits transferring annually between 1882 and 1907 – prompting him to conclude that the Cardwell/Childers reforms therefore benefitted the regulars more than they did the Militia.[29]

At least in this regard the reforms fulfilled their aims. The Volunteers, always more detached from the War Office due to their independent origins, lagged behind. The drive to reform the Volunteers had to continue under Lord Hartington's tenure in the War Office, with the introduction of the breech-loading Martini-Henry rifle in 1885 and the inclusion of an improved musketry qualification in the criteria for the capitation grant. However, according to Beckett, it was Edward Stanhope who 'did more to define a place for the Volunteers in national defence, and to develop Volunteer organisations accordingly, than any previous occupant of the War Office.'[30] Crucially, he integrated the Volunteers into the national mobilisation scheme, while placating Volunteer suspicions of overseas service by clearly stating

---

25   French, *Military Identities*, 20, 21–24.
26   Beckett, *Riflemen Form*, 134.
27   French, *Military Identities*, 216.
28   Ibid., 204.
29   Ibid., 214.
30   Beckett, *Riflemen Form*, 135.

that they would only be mobilised to resist an actual or apprehended invasion rather than a national emergency. Moreover, with agitation by Volunteers over the capitation grant and musketry qualifications rising, Stanhope established numerous committees to investigate where expenses could be saved by better management and relaxed the Volunteer musketry qualification somewhat.[31]

When seen in the context of the wider army reforms of this period, the changes in the Volunteers' organisation and effectiveness appear to have been more incremental than those in the army, or even the Militia. This was mainly due to the distinctive institutional origins and collective understandings of the Volunteers as a separate – but still related – entity to the army and Militia, most obviously manifested in its members' perception of the different function of their organisation: that of home defence. This position differed greatly from the actual functional requirements of a reserve organisation as defined by the state: that of a cheap method of quickly reinforcing the regular army in times of crisis. Indeed, it is possible to argue that this distinction between perceived function and required function of the army reserves resulted in a process of serial incrementalism rather than a single major transformational event during the Cardwell/Childers era. Similarly, although there was some increased co-operation, it is clear that the full integration of the Volunteers and regulars envisaged by Cardwell failed to materialise, mainly due to the different organisational nature of the Volunteers.

While Childers' reforms did conclude Cardwell's transformation of the army, crucially, the Militia – and to an even greater extent the Volunteers – lagged behind. For example, the Second Boer War of 1899–1902 raised serious questions about the effectiveness of the Militia, while highlighting disorganisation in the Yeomanry and the lack of seriousness in the ranks of the Volunteers.[32] Most worryingly, although over 50,000 served in the war,[33] French and Walker argue they failed to backfill the army in the numbers required due to the voluntary nature of their service overseas. This underscored that the reserves were not able to meet the functional demands placed on it by an army engaged in expeditionary warfare.[34] Ultimately, while the reforms did succeed in turning the Militia into a draft finding body for the army, the quality of recruit remained poor, and only 8.5 per cent of Volunteers served overseas during the Boer War, a disappointing figure given the extent of patriotic feeling at the time. By the first decade of the 20th century, with the threat of European war mounting coupled with the wider army's poor performance in South Africa, it was increasingly obvious that these shortcomings would need to be addressed. That task would fall to Richard Burdon Haldane.

---

31 Ibid., 135–36.
32 Walker, *Reserve Forces*, 15.
33 Personal correspondence, Professor Vincent Connelly, 9.
34 French, *Military Identities*, 28; Dennis, P. (1987) *The Territorial Army 1907–1940*, Suffolk: Royal Historical Society, 8.

Haldane and the Territorial Force

Haldane's efforts to reform the army after the failures of the two previous Secretaries of State for War, St John Brodrick and Hugh Arnold-Forster, have been well documented.[35] An intriguing and controversial character, opinion is also split as to Haldane's legacy.[36] Despite this attention, few works examine in detail the impact of his reforms on the various British reserve forces. Given Europe's rising militarism, the strategic uncertainty of the early 1900s, and Britain's continued Indian commitment, it is perhaps surprising that, like Cardwell, Haldane undertook his reform of the Volunteers for primarily economic rather than strategic reasons. As Spiers has noted 'it was the economy and not [the strategic situation in] Europe that had been the *sine qua non* of Haldane's army reform.'[37] These reforms occurred in a context similar to Cardwell's, with increasing political attention focused on the cost and effectiveness of the army and the reserves after an expensive war had once more highlighted their inefficiencies. Indeed, such was the growing political demand for change in the wake of both the army's and the reserves' poor Boer War performance, that the Conservative war minister Arnold-Forster had attempted to reverse linking altogether and create larger depots to provide recruits for all regiments while at the same time cutting costs, in what would have been a predecessor of today's centralised super-garrisons. While Arnold-Forster's attempts to reform the army and the reserves failed due to large and sustained resistance in Parliament, and in particular from the Army Council,[38] his efforts did pave the way for Haldane's reforms, allowing the new Liberal Secretary of State for War to emphasise the continuity of his policies with those of the Conservatives. Decisively, Haldane cemented cross-party political support for his reforms by assuaging the Radicals' fear of militarism while highlighting to the Tories how much they would save the War Office as well.[39] This economic argument was crucial, as a recent Royal Commission to investigate constant over-expenditure on

35   Spiers, E. (1980) *Haldane: An Army Reformer*, Edinburgh: Edinburgh University Press, 4–7; Satre, L.J. (1976) 'St. John Brodrick and Army Reform 1901–1903', *Journal of British Studies*, 15(2); Haldane, R. (1929) *Richard Burdon Haldane. An Autobiography*, London: Hodder and Stoughton; Koss, S. (1969), *Lord Haldane: Scapegoat for Liberalism*, New York: Columbia University Press; Spiers, 'The Late Victorian Army'; Higgens, S. (2010), 'How was Richard Haldane able to reform the British Army?', Unpublished MPhil. dissertation, University of Birmingham.

36   Fraser, P. (1973) *Lord Esher: A Political Biography*, London: Hart-Davis; Barnett, C. (1980) 'Radical Reform 1902–14' in Perlmutter, A. and Bennett, V. (eds) *The Political Influence of the Military: A Comparative Reader*, New Haven: Yale University Press: Spiers, *Haldane*; Higgens, 'How was Richard Haldane able to reform the British Army?'.

37   Spiers, *Haldane*, 73.

38   Beckett, *Riflemen Form*, 236; Dennis, *The Territorial Army*, 4.

39   Morris, A. (1971) 'Haldane's Army Reforms 1906–8: The Deception of the Radicals', *History*, 56(186).

the army and the resulting Treasury-imposed cutbacks eloquently concluded that 'extravagance controlled by stinginess is not likely to result in either economy or efficiency.'[40] Reducing the army by 20,000 men and decreasing the total number of active Militia, Volunteers and Yeomanry from 364,000 to 300,000 would allow this cycle to be broken, and the savings could be re-invested into re-structuring the Volunteers into a new, army-controlled Territorial Force (TF, renamed the Territorial Army in 1920).[41] This reform alone would reduce the reserves budget from over £4.4 million to £2.89 million per annum (the equivalent of a £140 million saving today),[42] and, for the first time in years, bring the entire army budget in below the £28 million (£2.6 billion today) ceiling allocated to it.[43] By stressing the substantial savings to be made, Haldane was also able to gain Liberal support while simultaneously outmanoeuvring opposition in the Army Council. Thus, once again, the reform of the reserves was undertaken for primarily economic reasons and was instigated by politicians rather than generals.

Yet, like the Cardwell reforms, Haldane's economic arguments did not occur in a political vacuum, and the subsequent organic development of the TF was undertaken in propitious circumstances conducive to the fusion of economy with ideological argument. The Liberals had been elected in 1905 on the platform of 'Peace, Retrenchment, and Reform' and were thus ideologically predisposed to the radical reform of the army and reserves that was clearly needed after the Boer War. Haldane himself was heavily influenced by German philosophy, and his vision of a new 'Hegelian army' that reconciled the military need for defence with the political need for economy was the ideological cornerstone on which his reforms rested.[44] Supporting his reforms, he declared that: 'The basis of our whole military fabric must be the development of the idea of a real national army, formed by the people, and managed by specially organised local associations.'[45] Echoing Gladstone, Haldane even suggested that the new TF would 'become a military school for the nation', indicating his hope that a reinvigorated reserve would both attract much-needed recruits and have an important moral impact on society.[46]

Strategic arguments were also deployed to gain support for the reforms. The 1903 Nicholson Commission had concluded that the threat of invasion from the Continent had declined significantly and that a smaller reserve force was therefore required for home defence. However, reserve forces would need to be more reliable

---

40 Spiers, 'The Late Victorian Army', 197.

41 Beckett, *Riflemen Form*, 235–38.

42 Ibid., 249.

43 Dennis, *The Territorial Army*, 8.

44 Vincent, A. (2007) 'German Philosophy and British Public Policy: Richard Burdon Haldane in Theory and Practice', *Journal of the History of Ideas*, 68(1); Howard, M. (1967) *Lord Haldane and the Territorial Army*, London: Birkbeck College, 10, 15.

45 Spiers, E. (1980) *The Army and Society, 1815–1914*, London: Longman, 268.

46 Beckett *Riflemen Form*, 236; Dennis, *The Territorial Army*, 11; Spiers, 'The Late Victorian Army', 206.

and flexible to be capable of quickly reinforcing the army's new 'Expeditionary Force', which, it was increasingly foreseen, would serve on mainland Europe.[47] This, at least temporarily, resolved the 'blue water' versus the 'bolt from the blue' strategic debate over whether the British military should place emphasis on expeditionary warfare or home defence.[48] In adopting such a strategy, a striking force of three army corps reinforced by elements of the TF was envisaged. Decisively, the consolidated TF was therefore to be the primary organisation by which the army could rapidly expand in times of need. By freeing the regulars of home defence duties and creating a decentralised TF administered on a local basis, Haldane hoped that he would create 'a British version of a nation-in-arms based on voluntary service' that would fulfil this role.[49]

In essence, Haldane's reforms were based on the central desire to ensure the largest possible expeditionary army that could be provided for during peace, while simultaneously reorganising the reserves into a single force that could reinforce the army when deployed overseas. Thus, he planned a regular Expeditionary Force 100,000 strong. This would be complemented by a Territorial Force of 300,000 incorporating the Militia, Yeomanry and Volunteers,[50] organised into 42 infantry brigades, 14 cavalry brigades and supported by full logistics elements. The new TF would be more closely controlled by the War Office, which stipulated that service contracts would be regularised at four years; annual military camps would last 15 days; all members were now subject to full military law; and that the army would be responsible for overseeing all TF training and assessing their readiness. To provide better oversight, new, predominantly civilian and elected Territorial County Associations would be established across the country, responsible for raising and administering their local units, but under the central direction of the War Office. Crucially Haldane intended to create a reserve force that, following six months' training after initial mobilisation, would be ready to deploy overseas with the British Expeditionary Force.[51] Essentially, he was attempting to professionalise the reserves and create a two-tier military readiness force structure that once again would balance demands for economy with the need for strategic flexibility.

However, while his reforms had widespread backing in Parliament, Haldane was to meet heavy resistance from the reserve organisations themselves. Firstly, the Militia representatives' intransigence when given the choice of integration with the regular army or joining the new TF led to the failure to reach any agreement and resulted in Haldane abolishing the Militia altogether. Instead, he created a small Special Reserve to keep a flow of draftees willing to serve in the army in time

---

47   Beckett, *Riflemen Form*, 232–33.
48   Dennis, *The Territorial Army*, 5.
49   Spiers, 'The Late Victorian Army', 206.
50   Ibid., 211.
51   Beckett, *Riflemen Form*, 248.

of war.[52] While the Militia's disbandment and replacement simply formalised the reality of its function, Volunteer and Yeomanry resistance to the erosion of their autonomy by the new County Associations also caused Haldane to rethink parts of this policy. Faced with increasing opposition from commanding officers, he was forced to drastically reduce the elected membership of the Associations, effectively ceding control of the bodies to the Territorials themselves and undermining a central tenet of his policy. Thus, as Dennis has noted, for 'the price of minimising Volunteer intransigence, a key element of Haldane's concept of the National Army was sacrificed before the Territorials were even born.'[53] Worse was to follow.

Haldane had announced in Parliament that the TF would serve overseas in support of the regulars. But when he introduced the Territorial and Reserve Forces Act eight days later on 19 June 1907, this decisive clause had been dropped. The reason behind this omission from the Act was twofold. Firstly, the Volunteers and their 'trade union in the House of Commons'[54] had strenuously objected to the introduction of the overseas obligation and Haldane needed their support in order to man the TF. Given their opposition, Haldane risked a recruiting crisis if he did not allay their fear of overseas service. In bending to their demand to drop stipulated overseas service, he instead hoped that their voluntary ethos would see between a sixth and a quarter of the TF volunteer for service with the army abroad if need be.[55] Secondly, as Dennis has noted, the change was also aimed at placating the more radical critics of his reforms who saw the very creation of an Expeditionary Force itself as disturbing.[56] Introducing the Act, Haldane thus changed tactics, stating that the role of the Territorials was primarily home defence. This was at odds with the whole thrust of his reforms which had been to create the Expeditionary Force and a reserve to support it. Thus, with strong organisational resistance threatening to undo his plans, for political expediency Haldane sacrificed the most vital tenet of his reforms to ensure his new organisation was not still-born. His last-minute climb-down would have far-reaching implications for the TF's organisational development and performance over the next century, creating confusion as to what the exact function of the TF was and when it could and should be used. This was as much a result of the institutional realities of the organisation and different stakeholder positions as it was Haldane's unwillingness to see his policies flounder. But the lack of clarity represented the start of a difficult and continuing juxtaposition within the TF between the state's functional need for operational flexibility and the reserves' institutional need to recruit.

In the end, the Act passed through Parliament with little resistance, and the creation of the Territorials on 1 April 1908 was strongly supported by the

---

52 Dennis, *The Territorial Army*, 12; Spiers, 'The Late Victorian Army', 210–11.
53 Dennis, *The Territorial Army*, 13.
54 Beckett, *Riflemen Form*, 250.
55 Spiers, *Haldane*, 108–13.
56 Dennis, *The Territorial Army*, 14.

King and the Lord Lieutenants who were to chair the newly formed County Associations. But it was clear that Haldane's reforms had been decisively weakened. One historian has said of Haldane that: 'He spoke and wrote in his memoirs as though he created a New Army. All that he had done was to rechristen the Volunteers.'[57] This is a little unfair, as Haldane had created the British Expeditionary Force (BEF) (albeit not the means to reinforce it) and the new TF did offer a more streamlined organisational framework that was now far stronger in terms of its supporting services and equipment. However, after an initial rise in recruitment following Haldane's Act, the TF still failed to meet its targets, with numbers decreasing to 268,000 by June 1909.[58] By September 1913 this had dropped to 236,000 actives,[59] about 60,000 short of establishment, while only one third of the force had achieved its musketry qualifications and just seven per cent had signed up for overseas service.[60] By that time, 80 per cent of the force were not re-engaging after their four years' service, and although better pairing between regular and TF units was evident, cultural divisions remained acute, with French arguing that the 'regulars remained almost as reluctant to accept Haldane's new creation as their equals as they had the Volunteers.'[61] This was especially evident in the exclusion of Territorials and old Volunteers from the regulars' regimental clubs, indicating the limits of integration and pairing even within the wider regimental family. As such, Haldane's Hegelian vision of a nation-in-arms never fully materialised.

The First World War

The following years saw much debate arise from the confusion over the Territorials' primary role. Such was the malaise within the TF and the complicated statutory position of its members in relation to overseas service that by the outbreak of the First World War seven years later the organisation was effectively by-passed in the national mobilisation plan. While the new Secretary of State for War, Field Marshal Kitchener, somewhat cruelly articulated his distrust of the TF as a 'town clerks' army',[62] the failure of both the army or the Associations to draft expansion plans, and the need for a home defence force added to the TF's perceived weaknesses. As a result of these, in 1914 Kitchener did not attempt to mobilise the Territorials along Haldane's two-tier plan, instead offering volunteers from the TF the chance to serve with their units if initially 80 per cent (later 60 per cent) of their unit's

---

57   Wilkinson, S. (1933) *Thirty-Five Years: 1874–1909*, London: Constable, 368.
58   Beckett, *Territorials*, 37–38.
59   Spiers, 'The Late Victorian Army', 211.
60   Beckett, *Territorials*, 39–40; French, *Military Identities*, 222.
61   Ibid., 227.
62   Dennis, *The Territorial Army*, 53.

establishment signed the Imperial Service Obligation (ISO) to serve overseas.[63] Given the wave of patriotism at the time, many Territorial units entered the regular army whole scale in this way, and by February 1915 there were already 48 Territorial infantry battalions in Flanders. The small numbers of units that had taken the ISO prior to war were immediately ordered to replace regulars on colonial duties, again indicating their lower status and their perceived lack of combat readiness. Importantly, after May 1915 – when larger formations of TF units were deployed overseas – they lost the suffix 'Territorial', indicating their assimilation into the regulars. By the time voluntary enlistment ended and direct recruiting into the Territorials was suspended in December 1915, over 725,000 men had joined its ranks over the previous 18 months; almost half of all those recruited to Kitchener's New Army.[64]

Given the need for mass mobilisation to replace the casualties on the Western Front, the Territorials, like the Militia before them, had in one respect become a drafting body for the army. It is important to stress here that this occurred by a process of assimilation, not integration; the Territorials were simply subsumed into the regular army. As units came up to strength with volunteers they were designated first-line units. At this point those who had not taken the ISO would revert to the second line units being filled by new recruits, and by November 1914 when the first-line units began to deploy, a third line unit would be established. This system eventually provided 318 battalions and 23 infantry divisions of 'Territorials' for service overseas, with the performance of these units widely praised, especially after they had adapted to field conditions.[65] Such was the importance of the volunteer Territorial units in the early stages of the war that Field Marshal John French later stated: 'Without the assistance that the Territorials afforded between October, 1914 and June, 1915, it would have been impossible to hold the line in France and Belgium.'[66] Indeed, the sombre statistic that the Territorials took over 577,000 casualties in all theatres of the war highlights their centrality to Britain's war effort, representing over a quarter of the army's 2,365,000 dead and wounded.[67] Ironically, when the Territorials were reconstituted in 1921, it was to be the shared sacrifices and the hard-won recognition of their fighting capabilities that increased their integration with the army more than Haldane's reforms ever had.[68] It is therefore clear that the model of Territorial mobilisation during the war did not follow Haldane's vision which had been decisively undermined by the exclusion of the overseas pledge. Indeed, the use of the TF as a draft finding and training body

---

63 Beckett, *Territorials*, 53.
64 Ibid., 58.
65 Beckett, *Territorials*, 57.
66 Dennis, *The Territorial Army*, 34.
67 Beckett, *Territorials*, 76; The War Office (1922) *Statistics of the Military Effort of the British Empire During the Great War 1914–1920*, London: HMSO, 237.
68 French, *Military Identities*, 230.

that was ultimately assimilated into the regulars rather than deployed as a reserve in its own right underscores that Haldane's reforms did not succeed in the one decisive area that they were designed to. Allison's and Kaufmann's organisational arguments offer a strong explanation for why Haldane's transformation ultimately fell short, but the evidence from this period also highlights that reforming the army's reserve forces has historically taken longer and proved more difficult than in regular forces. Moreover, the impetus for reforming the reserves again arose from the fusion of economic, strategic and politico-ideological goals.

The Second World War

Following their strong performance in the First World War, the Territorials had both earned the respect of the regular army and found themselves better integrated with them due to the shared trials of combat. However, the dire economic situation in Britain soon led to decreased defence spending, epitomised in the Geddes cuts of 1922. The newly renamed TA, still suffering from an ill-defined role due to the national 'ten year rule' defence strategy and a continued reluctance to accept a peacetime overseas pledge, found themselves bearing the brunt of these cuts. Throughout the inter-war years efforts to introduce an overseas service liability for the TA were rebutted by hostile County Associations who still resented the way the Territorials had been by-passed during the war. Under-recruitment remained a chronic problem, the nadir coming in 1932 when the TA was only 128,000 strong out of an establishment of 216,000, while technological advances left its equipment obsolete. Meanwhile, oscillations in Britain's defence posture between appeasement and a continental strategy saw the TA's function switch from air defence to second-tier reinforcement and back to air defence, with the Territorials wavering between being at the periphery and the core of defence planning. As so often the case with the reserves, it was only the looming European war and a rising sense of national emergency that eventually defined a role for the TA and saw its numbers swell, especially after the 1938 Munich crisis caused the government to double the Territorial's establishment. When war came, the Armed Forces Act of September 1939 suspended Territorial service for the duration, resulting in the assimilation of the TA into the army in a similar way to that which occurred in September 1914. The manner in which the reserves were assimilated into the regular army demonstrated once again that in times of national emergency the government could not afford the luxury of allowing the reserve army to serve only at home. However, the same was true of the TA itself and the citizens who now flocked to join its colours; both saw issues over the 'pledge' as unimportant when compared to national survival. As had been the case in 1859–60 when the French invasion scare saw the Volunteers created, it was increased perceptions of strategic threat that saw recruitment into the TA rise dramatically in the late 1930s. But such a fusion of public support with the political will to fund the reserve was relatively rare outside of wartime conditions; it was only the threat of existential conflict that saw the Territorials designated a role, properly invested in, and fully manned. And once this had happened, the TA was simply subsumed into the army again anyway.

Nevertheless, there remained clear evidence of disdain for the TA in the regulars, embodied by a lack of promotion of TA officers and a distrust of the quality of training Territorial units had received. This was hardly their fault. The rapid expansion of the TA from 1938 onwards had once again left it lacking NCOs to train the force, and this expansion rested on the assumption that Territorial divisions would have at least eight months collective training before they were deployed. Under increasing threat from Germany, this later changed to six months, and in the event, three Territorial divisions arrived in France in early 1940 after only four months training. Indeed, eight of the 13 BEF divisions deployed in 1940 were originally Territorial formations, and three of these – the 12th, 23rd and 46th Divisions, who had been tasked with rear security and lacked supporting arms and services – were thrown into the line in the retreat to Dunkirk. Some units in these formations had only one week's training, while others had never fired some of their weapons, many of which lacked ammunition. The 12th and 23rd divisions took very heavy casualties and were ultimately destroyed, but not before winning respect from the regulars and Germans alike for their tenacity. Nevertheless, the very heavy losses suffered by TA divisions in the defeat in France led to the break-up of most of those formations that did escape to Britain.

The debacle in France prompted a re-organisation of the surviving army, including a re-appraisal of how best to use the TA. Beginning during the First World War, the continuing rapid mechanisation of the combat arms in the inter-war period had precipitated a steep decline in the ratio of combat troops to support and logistics troops. Simultaneously, the dominance of infantry amongst the combat arms had also dropped from 53 per cent to 31 per cent, as mechanisation brought with it an increasing desire for armour, artillery and other mechanised support forces. These shifts in required force structure meant that many TA infantry units had to re-role. While this usually happened at the battalion level – with units re-training as armoured, parachute, signals or artillery specialists – it also occurred at the divisional level, with the 52nd Division assigned as mountain warfare experts. Similarly, the logistics arms were also forced to rationalise to meet the demands of increasingly mechanised warfare. For example, in October 1942, the Royal Electrical and Mechanical Engineers were formed to rationalise vehicle repair, for which the Royal Engineers, Royal Army Ordinance Corps, and Royal Army Service Corps had all been responsible. As would be expected, advances in technology therefore shaped the strategic and tactical environment which shaped reserve organisation.

A policy of bolstering TA units with regulars was also introduced for both the combat and logistics arms. Thus, most original TA divisions had one regular combat battalion per brigade, while regular divisions also usually had one to three

TA combat battalions per division. Support and logistics arms could be drawn from both regular and TA in both divisions, and cross-posting was also common. The net result was that, when not used as piecemeal infantry units against armour, as in France, the assimilated Territorial units performed reasonably well at the Second Battle of El-Alamein, in Tunisia, during the Normandy battles, and the Western European campaign that followed. This was especially true if they had been exposed to combat incrementally. For the most part, the policy of combining Territorial and regular units in larger formations appears to have boosted combat and logistics performance, while cross-posting also meant that the distinction between a former TA soldier and a regular was lessened. This continued after the war, as National Service saw a constant rotation of ex-servicemen through the Territorials as part of their obligation, and the deployment of some volunteers to Libya and Aden. But for the main the TA reverted to its home defence role, and when the phasing out of National Service was announced in 1957, accompanied by the changed strategic priorities of the nuclear age and the uncertainties of the British economy, the Territorials again found themselves the target for reform.

Carver-Hackett Cuts Deep

The origins of the reforms undertaken by General John Hackett and Major General Michael Carver are to be found in the 1964 ascension to power of a Labour government committed to putting Britain's 'defences on a sound basis and to ensure the nation gets value for money.'[69] Slow GDP growth and the devaluation of the pound forced the new government to seek economies. Moreover, the Labour Party had been elected on the promise of more funding for social programmes without seeking more taxation, and, as the Liberals before them, they viewed defence as an area where savings could be made. The cost of maintaining a nuclear deterrent, contributing to NATO and maintaining significant military capacity overseas was argued to be overbearing, and while the withdrawal from east of Suez would provide some savings, it was in this context that an earlier Defence Review report had concluded the cost of the TA could not be justified.[70] Similarly, the strategic rationale for a reduction of the Territorials was also made by reference to the new nuclear environment and the prevailing NATO 'short war' scenario, both of which, it was argued, rendered a large home defence force redundant.

The subsequent 1965 Defence Review was traumatic for the Territorials. The reforms it envisaged were based on Carver's assumption that the sole function of the TA was to provide a means by which the regular army could expand in

---

69 French, *Military Identities*, 200.
70 HM Government (1965) *Sir Philip Allen's Home Defence Review Committee Report*, London: HMSO.

wartime, and that it was failing in this role.[71] As a result, he proposed a slashing of the annual Territorial budget from £38 million to £20 million (£380 million today) and a re-orientation away from combat arms to support services. Crucially, the reforms were heavily focused on logistics. Almost half of TA units were designated support formations, while a 1,500 strong force of high readiness logisticians was complemented by an expanded force of 11,000 to provide support for the strategic reserve.[72] However, the expansion of reserve logistics capability was offset by severe reductions in TA manpower, with its established strength cut from 200,000 to 64,000.[73] Much of these fell on the 'teeth arms'. A total of 73 infantry battalions, 41 artillery regiments and 19 armoured regiments – effectively meaning the end of the Yeomanry – were cut, leaving only 13 infantry battalions, four artillery regiments and a single armoured regiment.[74] Moreover, of the 59 County Associations, only 14 would remain, in a deep blow to those organisations that had administered the TA since its inception. With one fell chop, the system instituted by Haldane's reforms had been all but eliminated.

Not surprisingly, the proposed reforms faced considerable opposition, most notably from the Council of Territorial Associations which had not been consulted by Hackett prior to the 1965 White Paper and also, it emerged, which had had their proposal of a cyclical limited liability for teeth arm units rejected out of hand. Although there was no statutory requirement for the Councils to be informed, such was the army's desire to push through the TA reforms that Hackett remarked 'there is an erroneous impression to the extent to which the scheme is open to discussion'.[75] With Carver likewise warning that there would be 'no climate of change' around the reforms, negotiations between the Council and the Ministry of Defence quickly broke down. Meanwhile, the Conservatives, rallying to protect 'one of sacred cows of the Tory establishment',[76] were defeated in a parliamentary no-confidence vote on the reforms by a single vote. Such resistance did result in some concessions from the army and the Labour government; an extra 28,000 light infantry being authorised before the Reserve Forces Act came into effect in 1967. The Act also reorganised the TA and Special Reserve into a four-tier Territorial Army and Volunteer Reserve force held at different levels of readiness and liability; this was essentially a consolidation of previous arrangements. But even with this consolation and the consolidation, the Carver-Hackett reforms drastically weakened

---

71 Ibid.

72 Beckett, *Territorials*, 201.

73 Walker, *Reserve Forces*, 20; Stanhope, H. (1979) *The Soldiers: An Anatomy of the British Army*, London: Hamish Hamilton, 250, has the same figure. Beckett provides the number 107,000 to 57,000 without citation; this may represent actual trained strength rather than total establishment.

74 Beckett, *Territorials*, 201.

75 Ibid., 202.

76 Beckett, *Territorials*, 203; Stanhope, *Soldiers*, 251.

the TA and left its members and those on the surviving County Associations deeply suspicious of the top echelons of the army, whom they felt had betrayed them in order to save the regulars from the worst of the cuts.[77]

Once again, it is clear that primarily economic arguments were fused with those of ideology and strategy in the decision to undertake the reforms. However, what is interesting about the Carver-Hackett reforms is that they were comparatively successful at instigating reserve organisational transformation. What allowed transformation to be driven through to its conclusion was the personal determination of Carver and Hackett to instigate changes, and, most importantly, the political support of the government which was conducive to their and the MoD's unwillingness to negotiate with the Council. Although resistance from stakeholders was forthcoming, in comparison to Cardwell's and Haldane's more consensual approach, the 1967 reforms were to a large degree presented as a fait accompli by the army's elites and simply pushed through from the top down, with only one minor modification. However, such an approach not only caused long-lasting distrust between the TA and the regular army, it also saw the first time that reform of the TA became highly politicised, with the breakdown of cross-bench support for the reforms evident in the Conservatives' reaction. The army's lack of consultation with the Council was a major cause of this, but the Carver-Hackett reforms were important in that from now on the revival or reduction of the TA would become increasingly politicised along party lines. Indeed, it is noteworthy that all the previous periods of major reform were undertaken by Liberal governments; as one Tory aide has stated: 'The Conservative Party has liked the TA for two reasons: it fosters the volunteer ethic and it is very cost-effective.'[78] As such, when the Conservatives returned to government in 1970 it was no surprise that they quickly increased the Territorials establishment by 10,000, even though the force continued to suffer from chronic underinvestment and under-recruitment.

Following the publication of the Shapland Report in 1978, the TA experienced a revival under Margaret Thatcher's government, with numbers expanded to 86,000 on paper. The force also enjoyed an increase in investment under Thatcher and, by 1984, had a more clearly defined role: that of the rear defence of NATO areas of operation on the Continent. As a result, the Territorials averaged 89 per cent of established strength between 1979–89, but the 25–30 per cent annual soldier wastage rate remained a major problem.[79] Nevertheless, there were major concerns within the army about the TA's ability to meet this new role.[80] With cashing-in on the peace dividend a priority in the 1990 'Options For Change' programme, the TA escaped comparatively lightly, with a reduction to 63,500 somewhat offset by its inability to reach its full establishment anyway. These cuts fell heavily on TA

77   Walker, *Reserve Forces*, 21.
78   Ibid., 51.
79   Beckett, *Territorials*, 212.
80   Bennest in Walker, *Reserve Forces*, 76.

combat forces, with the programme instead prioritising reserve logistics capability, similar to Carver-Hackett. As combat forces were often the easiest to recruit, the resulting lack of balance created manning problems that were to last for decades Moreover, chronic neglect by the army remained a major problem, and structural problems due to the reduced size of both the army and the TA were not addressed. The 1996 Reserve Forces Act, amongst other measures, changed the call-out terms for reservists so that the Secretary of State for Defence, not Parliament, could mobilise reservists if need be. However, this once again highlighted the political question of when the reserves should be used, especially given the intensification of army operations abroad after Tony Blair became Prime Minister in 1997. Since that date, the TA continuously contributed 10–12 per cent of the UK's total mission force in both Iraq and Afghanistan, but these forces were predominantly deployed as individuals to backfill regular units, rather than as formed units. Moreover, apart from the initial invasion of Iraq in 2003 (TELIC 1) when there was a general mobilisation, during this period the Territorials were usually reliant on individuals to volunteer for service rather than compelling them to do so. This policy was known as 'intelligent mobilisation' and it reflected both a wariness amongst regular army leaders about TA quality, and also political sensitivities about extending mobilisation during limited and often unpopular wars abroad. However, it also severely limited the ability to deploy formed TA units overseas, an indication that Haldane's last-minute exclusion was still curtailing the utility of the organisation a century later.

Conclusion

What do these past periods of reserve reform tell us about these processes, and the army's reserve in general? Firstly, it is clear that attempts to reform the reserve are cyclical in that they occur in response to changed economic and strategic circumstances which provide the impetus for another cycle of reform. The primarily financial impetuses for reforming the reserves have been supported by the politico-ideology of those undertaking the reforms. Secondly, throughout their history the army's reserve forces have often come under pressure to reform after wars, and that reform is frequently, but not exclusively, attempted simultaneously with that of the army. Thirdly, the sources of reform have also been primarily located in the political rather than the military sphere, and where the army has been keen to implement reserve transformation, this has often been in the context of the struggle for organisational survival epitomised in reductions in military spending. Fourthly, the most recent reserve reforms have focused on cutting combat forces and maintaining logistics functions in order to save costs, but this has resulted in recruitment issues. Finally, as I have shown, apart from times of national emergency, the reserves have historically struggled to recruit to full strength. Indeed, Peter Caddick-Adams has noted that 'whatever the manpower establishment, the Territorials seem to hover at 10 per cent below', although it should be noted that some of this is usually explained by the way the TA has traditionally recorded its trained strength and establishment.[81]

However, the most important conclusion to be drawn from this chapter concerns the impact of the previous periods of reform, and in particular the time it took to effect organisational change within the reserves during each era. Throughout, stakeholder resistance and organisational friction within the army, the reserves and Parliament – most frequently caused by recruitment issues and potential deployment overseas – have consistently limited the impact of reforms. This fact highlights how these two issues are fundamental to understanding today's Army Reserve and the attempt to transform it. Indeed, it is clear that delays to transformation are inherently bound up in the organisational nature of a part-time force. Almost every period of reform has taken years to implement, much longer than originally intended. As one TA Colonel has remarked: 'There is nothing in the TA you can do immediately ... it takes five to ten years [to make changes].'[82] And when these changes have finally been implemented their impact has been generally more limited than originally envisaged; each reform has been adjusted due to political and organisational resistance. Indeed, most reforms have failed in their primary focus of making the army's reserve more operationally deployable. Thus, it appears that the part-time, volunteer and citizen nature of the reserves inherently limits transformations when compared with the regular army. British reserve forces at least have always been slower and more difficult to reform, and much of this has been related to their organisational resistance to be deployed overseas *en masse*. Drawing on this evidence, I would contend that in general, reserve transformations take longer to effect than those of regular forces due to the distinct character of their organisations. Indeed, it is possible to argue that the fundamental nature of Britain's army reserve has changed little over the past 150 years. Yet, as this chapter has shown, while the economic, ideological and strategic questions underpinning the Haldane, Cardwell-Childers, Carver-Hackett reforms – and the FR20 transformation discussed in the next chapter – are remarkably constant, their organisational outcomes were more a product of their own historical contexts than the similarities between these questions would suggest. While the rationale for reforming the reserves may often bear semblance to previous attempts, how transformation actually occurs is based firmly in current organisational realities. The questions may be the same, but the solutions are different.

Over its history, it appears that the reserves have changed by numerous processes identified in the military transformation literature outlined in Chapter 1.

---

81 Caddick-Adams, P. (2002) 'The Volunteers', in Alexandrou, A., Bartle, R. and Holmes R. (eds) *New People Strategies for the British Armed Forces*, London: Frank Cass, 95; Personal correspondence, Vincent Connelly July 2018.

82 Walker, *Reserve Forces*, 21.

While Cardwell's emulative attempt at organisational reform represented a top-down process of change in both the army and the reserves, it was severely hampered by stakeholder resistance, organisational friction and the changing strategic imperatives of a withdrawal from the colonies. The subsequent top-down Childers reforms represented an attempt to relieve the organisational pressures of recruitment and retention that resulted from Cardwell's re-structuring and the changed strategic circumstances. Following Allison, while both Cardwell and Childers provided the personal drive needed for change, the reserves in particular evolved as much by the process of incremental adaptation – itself caused by the friction associated with the struggle for organisational survival – as they did by top-down direction. The experience of the Territorials in the First World War also suggests that it was adaptation to battlefield realities in the field that honed the TF's skills and allowed better integration with the army, rather than the reform process that Haldane had instigated. Meanwhile, the ultimate trajectory of Haldane's reforms was shaped by both the external pressures of economy, strategy and ideology, but most importantly, by the internal need to recruit the Territorials. Indeed, it was this functional requirement and the debates that resulted from it that ultimately shaped the development of the Territorials during this period. The Carver-Hackett reforms were noteworthy for the manner in which top-down transformation was imposed upon the TA without cross-party political consensus, and driven through by an army keen to protect its own organisation. The generals' unwillingness to compromise caused wider political fallout, including lasting distrust between the reserves and the army. This has not been helped by the latter's neglect of the reserves during peacetime, and then their assimilation during wartime.

Chapter 3

The Transformation of Military Logistics

So far, I have examined the historical cases of reserve reform in order to contextualise the current transformation. However, it will also be remembered that FR20 placed heavy emphasis on the domain of logistics. As the policy stated 'Greater reliance will be placed on the Reserves to provide routine capability ... primarily in the areas of ... combat service support (such as logistics ...)', it is therefore necessary to understand how military logistics have developed in the 21st century.[1] Indeed, this attempt to transform the reserves and especially its logistics component cannot be understood without recognising the drastic changes in how logistics is delivered. Consequently, it is necessary to discuss the evolution of logistics in significant detail before examining the specific new methods of organising and delivering logistics in the reserves. Thus, in this chapter I explore the wider processes driving the recent development of Western military logistics in order to understand the organisational context for FR20, and ultimately, the reserves logistics sub-units examined later in this study. It is important to situate FR20 within the wider post-Fordist conceptual approach to military logistics as the policy vastly increased the capability requirements expected of reserve logistics sub-units, and hence the need for greater professionalism and cohesion within them. Furthermore, the post-Fordist approach to logistics not only immediately affects the sub-units examined in this study through the processes identified in this chapter, it also provides a historically novel solution to the recurrent organisational problems experienced in past periods of reserve reform I outlined in the last chapter.

In this chapter, I make two arguments. First, I draw on the existing post-Fordist literature to provide a conceptual framework for understanding the recent transformation of Western military logistics. In doing so, I challenge the classical literature on military logistics to show how much of what has been written on the subject is either out of date and lacking sociological theory. Second, I combine this post-Fordist framework with the transformation literature to show how these changes have occurred through the simultaneous processes of centralisation;

---

1 *Future Reserves 2020*, 22.

supply chain management and outsourcing; using core and periphery forces; and by adopting a networked approach to logistics. Throughout, evidence is provided from US, British and NATO military logistics structures and practices. The relevance of the US' adoption of a post-Fordist approach to logistics is critical. As I will show, it has predominantly led the way in this regard, subsequently influencing both the UK and NATO to follow their 'best practice'. The conclusion argues that a logistics transformation has occurred, is ongoing, and that post-Fordism is a useful conceptual framework to understand it. Indeed, it argues that such is the scope and nature of this transformation that it has had a profound impact on wider Western force structures, and ultimately, their strategic flexibility.

Since the end of the Cold War, numerous authors have identified a major shift in the nature of modern conflict and a transformation in the organisation of Western military forces.[2] Replacing inter-state conventional conflicts, insurgencies, proxy and civil wars, and terrorism have come to dominate the character of the 'new wars' the West has fought. Meanwhile, enabled by the advanced technology of the Revolution in Military Affairs (RMA), smaller, more professional, capable, and agile combat forces are orienting themselves toward a more diverse set of expeditionary missions in more operationally challenging locations around the globe.[3] These trends have been accompanied by the simultaneous growth in the privatisation of Western militaries.[4] While debate remains over the extent of some of these changes,[5] in almost all cases, the focus of the works that examine these military transformations has been on combat forces.

Despite this focus on combat forces, strategists, military commanders, and theorists throughout the ages have all remarked on the importance of logistics in successful combat operations and, ultimately, in implementing strategy.[6] As General Omar Bradley's oft-cited quote that 'Amateurs study strategy and tactics, professionals study logistics' highlights, in the professional military, combat commanders view the study of logistics as fundamental to the success of operational

---

2   Kaldor, M. (1999) *New and Old Wars: Organised Violence in the Global Era*, Cambridge: Polity; Duffield, M. (2001) *Global Governance and the New Wars*, London: Zed Books; Smith, R. (2005) *The Utility of Force*, London: Allen Lane.

3   King, *The Transformation of Europe's Armed Forces*; Demchak, C. (2003) 'Creating the Enemy: Global Diffusion of the Information Technology-Based Military Model', in Goldman, E. and Eliason, L. (eds) *The Diffusion of Military Technology and Ideas*, Stanford: Stanford University Press; King, A. (2009) 'The Special Air Service and the Concentration of Military Power', *Armed Forces and Society*, 35(4).

4   Avant, D. (2005) *The Market for Force*, Cambridge: Cambridge University Press; Kinsey, C. (2005) 'Regulation and Control of Private Military Companies', *Contemporary Security Policy*, 26(1); Krahmann, E. (2005) 'Security Governance and the Private Military Industry in Europe and North America', *Conflict, Security and Development*, 5(2).

5   Berdal, M. (2003) 'How "New" are "New Wars"? Global economic change and the study of civil war', *Global Governance*, 9(4).

6   Kane, T. (2001) *Military Logistics and Strategic Performance*, London: Routledge, xiv.

plans.[7] Those academic works that exist on contemporary military logistics note that 'dramatic change' has occurred in the past 20 years; they also state that this change remains understudied.[8] Moreover, much like the RMA, it is important to situate the start of the British transformation of military logistics within its US context. As I show in this chapter, the British military has consistently mirrored the US example in terms of adopting new logistics concepts and doctrine. Similarly, it is also important to understand that this logistics transformation is not unique to the British and US militaries; NATO, and non-NATO states are also adopting post-Fordist approaches to organise their logistics and wider force structures.

It is important to note that while post-Fordism is an analytical term that academics have coined to describe the changes happening in some Western militaries since the 1990s, neither the wider UK defence establishment, nor the British Army in particular, use the term. Nor should they. No military officer would consciously describe themselves as a 'post-Fordist'. Indeed, during the research for this book, when I used the term, I was directly asked what this meant by both a former Chief of the General Staff (CGS), and by a Colonel responsible for implementing future logistics doctrine and concepts.[9] Nevertheless, when I explained the four tenets of post-Fordism, both immediately concurred it was an accurate term to describe the processes ongoing in the army and its logistics component. Thus, while the terms post-Fordism and post-Fordist principles are used throughout this work, it must be stressed that although the army has implemented these four central tenets, it has not done so self-consciously. More, as I show in this chapter, they have emulated business best practice in doing so. As such, I provide evidence of the processes of post-Fordism within British military logistics rather than argue that the British military has consciously implemented a post-Fordist logistics 'model'.

Military Logistics

Despite numerous military and academic definitions of military logistics,[10] I contend that a simple unifying principle unites these definitions: *support to military*

7   Pierce, T. (no date) 'US Naval Institute Proceedings', 122(9), 74.

8   Rutner, S.M., Aviles, M. and Cox, S. (2012) 'Logistics evolution: a comparison of military and commercial logistics thought', *The International Journal of Logistics Management*, 23(1), 96; Erbel, M. and Kinsey, C. (2018) 'Think Again – Supplying War: Re-appraising Military Logistics and Its Centrality to Strategy and War' *Journal of Strategic Studies*, 40(4), 1.

9   Personal communication, General Sir Peter Wall, 10 May 2016; Interview, senior British Army logistics officers responsible for future doctrine, 9 June 2015.

10 Uttley, M. and Kinsey, C. (2012) 'The Role of Logistics in War', in Strachan, H. (ed.) *The Oxford Handbook of War*, Oxford: Oxford University Press; Van Creveld, M. (2009) *Supplying War*, Cambridge: Cambridge University Press, 1; Lynn, J. (1993) *Feeding Mars: Logistics in Western Warfare from the Middle Ages to the Present*, Boulder: Westview Press, ix; Kane, *Military Logistics*, 2; U.S. Department of Defense (2013) *Joint Warfare Publication*

forces synchronised through space and time. In short, military logistics from the ancient to the modern era has always been about *getting the required quantity and quality of material and services, to the correct place, at the correct time, and in the correct order, to ensure military forces are as capable as possible.* This definition is used here. At the same time, I acknowledge that this chapter cannot address each and every element of military logistics contained above, and will instead focus on specific areas where British, US and NATO logistics processes and practices have been transformed.

Although the last decade has witnessed intense military activity, and at times strong media focus on Western military logistics failures, it is only very recently that military logistics has begun to receive scholarly attention. This can be broadly categorised as works concerning how logistical constraints shape military strategy and operations; those on national logistics, and those detailing some of the rapid and complex changes occurring in the sector. Michael O'Hanlon's work on logistics constraints on US force projection and Eugenio Cusumano's work on contractor support in the US and UK are examples of the first group.[11] In the second, John Louth and David Shouesmith have provided detailed analyses of the complex problems within current British defence logistics.[12] Importantly, Shouesmith has noted how this complexity reflects changes in the nature of the modern state.[13] Mikkel Rasmussen has argued for closer integration of civilian business strategy and organisation in Western militaries to address this complexity.[14] The third category is divided in two. Matthew Uttley and Christopher Kinsey argue that 'the inherent nature of defence logistics ... has remained constant since the era of ancient warfare' and that 'the steps required to construct and operate a logistics system have remained conceptually simple and timeless'.[15] Conversely, Mark Erbel and Christopher Kinsey argue that a Revolution in Military Logistics (RML) has occurred of such importance that without it the RMA would have been still-born. But, although their article suggested a new conceptual approach to military logistics, they failed to provide a theoretical framework to explain the processes behind the RML. While collectively these works indicate that scholars are converging on the topic, I contend that they lack a conceptual framework

---

4-0: *Logistics*, para GL-7; MoD (2015) *Joint Doctrine Publication 4-00: Logistics for joint operations (fourth edition)*, London: HMSO, x.

11 O'Hanlon, M. (2009) *The Science of War*, Princeton: Princeton University Press; Cusumano, E. (2016) 'Bridging the Gap: Mobilisation Constraints and Contractor Support to US and UK Military Operations', *Journal of Strategic Studies*, 39(1).

12 Louth, J. (2015) 'Logistics as a Force Enabler: The Future Operational Imperative', *RUSI Journal*, 160(3).

13 Shouesmith, D. (2001) 'Logistics and Support to Expeditionary Operations', *RUSI Defence Systems*, 14(1).

14 Rasmussen, M. (2015) *The Military's Business: Designing Military Power for the Future*, Cambridge: Cambridge University Press.

15 Uttley and Kinsey, 'The Role of Logistics in War'; 402, 405.

for explaining the recent changes in military logistics and make no attempt to explain the wider processes that have led to these changes.

As a result, and despite earlier works on military logistics,[16] Van Creveld's 1977 *Supplying War* is still viewed as the seminal work on the subject. It is necessary to discuss this in detail here as it establishes the basis of my argument on post-Fordist theory's relevance to military logistics. In it, Van Creveld investigates logistics in predominantly European land campaigns from the 16th century to the 1944 Allied invasion of Normandy. For Van Creveld, the method of supplying armies during the period 1560–1715 was essentially feudal, based on Ancien Regime society, and in arguing that the military logistics system could not change until society changed, Van Creveld implicitly acknowledges that military logistics is fundamentally related to modes of production.[17] This chimes with other scholars who have examined how methods of supplying armies influenced the development of the European state.[18]

However, Van Creveld then challenges the centrality of dominant modes of production to understanding effective military logistics organisation and systems. In his most controversial chapter examining the highly detailed, synchronised and sophisticated Allied logistics plan for Operation Overlord, Van Creveld argues that it did not survive contact with the beaches, and that improvisation was again resorted to.[19] Indeed, for Van Creveld, improvisation defines successful military logistics. Despite an acknowledgement of the need for preparation, Van Creveld's central and counter-intuitive argument is that the impact of preparation on operations is limited and does not always equal success. He states that flexibility, resourcefulness and determination can overcome logistics weaknesses, and in doing so, he argues that continuity – in the form of logistical improvisation – is the defining characteristic of military logistics through the ages.[20] Decisively, Van Creveld remains unconvinced that systematic improvement in military logistics is possible as 'the results of the only comprehensive effort which was made in this direction [were not] particularly encouraging.'[21] This is a contentious position. While it downplays the importance of military logistics planning and systems in military outcomes, it also contradicts the opinions of many modern commanders on the importance of sound logistics preparations.[22] It can also be argued that Van

---

16 Thorpe, G. (2012) *Pure Logistics: The Science of War Preparation*, Charleston: Nabu Press.
17 Van Creveld, M. (1977) *Supplying War: Logistics from Wallenstein to Patton*, Cambridge: Cambridge University Press, 21.
18 Tilly, C. (1990) *Coercion, Capital, and European States, AD 990–1990*, Cambridge: Blackwell.
19 Van Creveld, *Supplying War*, 209.
20 Ibid., 203.
21 Ibid., 236; Kane makes a similar point in *Military Logistics*, 7.
22 Pagonis, Lieutenant General W. (1992) *Moving Mountains*, Boston: Harvard Business School Press; Smith, Lt Gen. R. (1995) 'The Commander's Role' in White, Maj. Gen. M. (ed.) *Gulf Logistics: Blackadder's War*, London: Brassey's.

Creveld's reasoning displays academic nihilism, for if improvisation is decisive, what is the point of studying military logistics organisation and systems?

Despite its original contribution, *Supplying War* has been critiqued by numerous authors. Thomas Kane has launched a sustained challenge to Van Creveld's suggestion that logistics preparations are 'futile' by examining campaigns from the Second World War to the onset of the RMA.[23] In all these cases, Kane details how careful attention to logistics planning and execution acted not only as an operational force multiplier, but also how such preparations gave military forces better strategic choices that ultimately allowed them to undermine their adversaries.[24] For Kane, 'logistics is the arbiter of opportunity … supply preparations not only help determine the character of a war, they are affected by the outcome of that determination.'[25] Thus, logistics organisation and military strategy are inherently interlinked; cause and effect can flow both ways. In challenging *Supplying War*'s final assertion that the human intellect cannot fully understand war and thus strategy, Kane argues that not only are logistics organisation and systems often decisive to military effectiveness, but critically, understanding military logistics is the first step toward understanding an adversaries' intent.

Both authors attempt to address the potential impact of information technology on military logistics. In a 2004 post-script, Van Creveld correctly identifies how computerisation and JIT logistics will allow the fine-tuning of logistics capabilities with operational needs, and how modern armies negotiate contracts for services on the free market. However, he then lazily concludes that there has been no fundamental shift in military logistics since the Second World War as the main method of supply is still predominantly based on road transport and intensive industrial modes of production. Crucially, Van Creveld states that: 'It does not appear as if the nature of logistics has undergone or is about to undergo a fundamental change'.[26] Kane also stresses continuity over change in rebutting claims that better information technology will 'obviate much of [the need for] logistics'.[27]

Van Creveld's position that logistical improvisation, determination and flexibility are decisive certainly has merit: flexibility remains a principle of logistics in NATO militaries. Van Creveld is also correct that most supplies are still shipped by road. Meanwhile, Kane, and Erbel and Kinsey are correct that logistics organisation affects strategic performance. And all these authors are correct in identifying that information technology is and will shape future logistics. However, these works lack theoretical depth in discussing the RML. Van Creveld and Kane lack a conceptual framework for understanding the RML and do not attempt to link it with recent changes in modes of production. Indeed, Van Creveld explicitly states

---

23   Kane *Military Logistics*, 7.
24   Ibid., 111, 171.
25   Ibid., 178, 10.
26   Ibid., 258.
27   Libicki, M. (1996) 'The Emerging Primacy of Information', *Orbis*, 40(2), 261.

that he is unconcerned with 'any abstract theorising'.[28] As a result, his emphasis on improvisation ignores the 'quiet revolution' in Western logistics organisation and systems over the last twenty years. His argument that the modes of production are the same as they were in 1944 is simply incorrect. While Kane's assessment of the impact of technology on logistics is more considered, it leaves open the question of what has changed in the 18 years since his work was published and how this change has occurred. Similarly, Uttley and Kinsey's arguments about the enduring nature of military logistics principles have been superseded by events: the US, UK and NATO have all since updated their principles in line with the post-Fordist approach.[29] Perhaps it is time to examine these changes.

Post-Fordist Industrial Logistics

In order to understand the dramatic change in how military logistics is delivered since the Cold War, and how this has shaped wider force structures, it is important to firstly examine industrial logistics. This is vital to highlight how the relationship between military and industrial logistics thought has changed over time, with industry in the ascendancy in the current era, to establish the conceptual foundation for the post-Fordist framework. In charting the evolution of logistics thought, Stephen Rutner et al. posit that while the practice of logistics originated in the military, 'civilian logistics and supply chain management surpassed military logistics at some point after World War II.'[30] This view is supported by John Kent and Daniel Flint, who examined the industrial logistics literature to describe its evolution in six key phases. The first phase is farm to market logistics which describes the transfer of goods from point of production to point of sale. By the start of the Second World War, Kent and Flint argue that this era had been largely eclipsed by 'segmented functions' logistics. The primary focus at this time was on the functions that distributed goods, with heavy emphasis on in-bound out-bound transportation, warehousing, wholesaling and inventory control, coupled with a reliance on the combustion engine to produce greater efficiencies.[31] Steven Simon describes this as the Fordist logistics model, based on a static supply chain 'in which the manufacturer contracts with a supplier to make and deliver material to the facility, where it is stockpiled.'[32] Kent and Flint argue that this era was heavily influenced by the military logistics practices of the Second World War that continued to be utilised by industry until the early 1960s. Similarly, Rutner

28 Van Creveld, *Supplying War*, 3.
29 U.S. Army (2012) *Doctrine Publication 4-0 Sustainment*; Ministry of Defence, *Joint Doctrine Publication 4.00: Logistics for Joint Operations*.
30 Rutner et al., 'Logistics evolution', 97.
31 Kent, J. and Flint, D. (1997) 'Perspectives on the Evolution of Logistics Thought', *Journal of Business Logistics*, 18(2), 23.
32 Simon, S. (2001) 'The Art of Military Logistics', *Communications of the ACM*, 44(6), 63.

et al. see the US Army's use of rear logistics bases in the European theatre (which Van Creveld argued were inefficient) as the 'precursor to the modern distribution centres used by the world's largest firms.'[33] For these authors, the military provided the impetus for change in industrial logistics, highlighting how military logistics was at the vanguard of logistics thought during this period. Crucially, the main body of Van Creveld's analysis of military logistics is based on evidence from this era and as a result, I contend this is where the utility of much of his contribution ends.

However, Kent and Flint argue that the era of segmented functions was followed by the development of 'integrated functions' in the early 1960s. This describes the trend toward viewing independent logistics functions holistically as part of a wider, interdependent system. During this period, as the business environment became more dynamic and competitive, there was a shift in emphasis from physical distribution to a 'total cost' approach to all parts of the logistics process, with a growing emphasis on information systems, services, marketing, and a wider realisation that one size of product did not fit all. This era coincided with the beginnings of post-Fordist modes of production, and these developments were advanced during the subsequent era Kent and Flint term 'customer focus' in the 1970s and 1980s. This involved a shift in primary focus toward the end user of the product, and toward maximising profits rather than minimising costs. Link node concepts of logistics, and greater emphasis on operations management and management science also emerged during this era.

With the onset of the eras of 'logistics as differentiator' and 'behaviour and boundary spanning logistics' in the 1980s, the relationship between industrial logistics and military logistics inverted, driven primarily by new information technologies. The realisation that information technology could support highly synchronised JIT logistics systems to increase commercial returns first originated in the production practices instigated by Toyota during the mid-late 1970s in response to inflation and a stagnating Japanese economy. Reducing waste – in the form of stocks, workforce, and production times – was the crucial motivator for the introduction of these practices. The basic premise of JIT holds that 'no product should be made, no component ordered, until there is a downstream requirement', and one of the central tenets of JIT logistics is Supply Chain Management (SCM).[34] SCM views the procurement, supply and distribution functions as a single system, and aims to 'establish control of end-to-end process in order to create a seamless flow of goods.'[35] Utilising information technology to increase control of the total supply chain reduces costs and increases profitability.

---

33 Rutner et al., 'Logistics evolution', 102.

34 Christopher, M. (1998) *Logistics and Supply Chain Management*, London: Pitman, 179.

35 Christopher, M. and Holweg, M. (2011) '"Supply Chain 2:0": managing supply chains in the era of turbulence', *International Journal of Physical Distribution and Logistics Management*, 41(1), 63, 69.

And with better control, the supply chain is more flexible to respond to changes in supply or demand. Decisively, however, the SCM approach is based on stable assumptions of demand and supply that were a product of the relatively stable strategic and market environment during the Cold War. As a result, SCM systems are dynamically flexible, but 'only within the set structure of their existing supply chain design.'[36]

Coupled with a greater understanding of the benefits of inter-organisational efficiency and reverse logistics within an increasingly globalised economy, SCM's cross-functional approach was central to the new JIT logistics procedures that were adopted by other Japanese and US firms in the early 1980s. Meanwhile, changing customer demands encouraged outsourced production and services to allow firms to respond to market demands.[37] Rutner et al. have also identified how the deregulation of the transportation industry in the US and a growth in mergers of US firms began the trend toward decentralised organisational structures and flatter hierarchies in wider industry. At the same time, logistics became central to production operations; streamlined and efficient logistics systems were seen as decisive in conferring competitive advantage. Kent and Flint show that recognition of this principle continues to grow today, and that an understanding of the benefits of co-operation between firms is leading to greater inter-firm and inter-functional co-operation and coordination of logistics efforts to increase both efficiencies and flexibility.[38] The underlying motivation for all these changes is that strategic alliances across the entire supply chain allows organisations to better adjust to changing customer demand while limiting costs.

It is therefore clear that modern business logistics has transformed in the last 30 years as the global economy and modes of production have evolved. It is also clear that the nature of the isomorphic relationship between military and industrial logistics has shifted since the Second World War. Moreover, as Rutner et al. and Flint and Kent show, there is an identifiable time-lag between the introduction of new business logistics practices, their appearance in industrial logistics publications, and then their adoption by military logisticians.[39] Thus, since the 1960s Western militaries, isolated from the industrial world and protected from market competition, have been slower to change their logistics practices than commercial firms. While this fact is important in explaining the beginnings of Western military logistics transformation, it should be noted that the different nature of military logistics – where military effectiveness is the ultimate standard by which success is judged, not efficiency – also played an important part in slowing the introduction of post-

---

36   Christopher and Holweg, 'Supply Chain 2:0', 64.
37   Mentzer, J., Soonhong, M. and Bobbit, M. (2004) 'Toward a Unified Theory of Logistics', *International Journal of Physical Distribution and Logistics Management*, 34(8), 620.
38   Kent and Flint, 'Perspectives on the Evolution', 26.
39   Rutner et al. 'Logistics evolution', 98, 102; Kent and Flint, 'Perspectives on the Evolution', 24.

Fordist techniques, as did stockpiling to increase resilience. However, in time these came to be seen detrimental to military effectiveness and efficiency.

Adaptation, Innovation, and the Legacy of Cold War Military Logistics

During the Cold War, NATO's strategy for deterring Warsaw Pact forces was based around the positional defence of Central Europe. This saw a significant proportion of combat forces based in forward positions in West Germany in order to fight a conventional, high intensity defensive war. These forces were to be supported by reinforcements held at varying degrees of readiness moving to predetermined positions in a 'layer cake' defensive plan, as shown in Figure 3.1. Crucially, each nation was responsible for logistics in their own sector, and each sector utilised linear lines of supply. To sustain such large, forward-positioned forces, NATO accepted stock levels were for 30 days of combat supplies. As a result, formations such as 1 British Corps organised their logistics at successive levels using the traditional 'echelon system', with stores held at frontline units, then forward storage sites, then at rear depots and finally larger quantities held in storage in locations such as Antwerp.[40] This structure meant that in the event of hostilities, the main logistics plan was based around the forward movement of stocks, with combat forces' controlled withdrawal along predetermined lines of communication gradually reducing supply lines. NATO's strong understanding of the Warsaw Pact's doctrine and tactics thus shaped its pre-determined defensive plan and its logistics plan. Moving pre-arranged levels of stock forward at pre-arranged times along secure lines of communication in rear areas (or indeed simply waiting for forces to withdraw to them) meant that there was little need nor desire for complicated asset tracking or inventory management systems, while the logistics structure itself remained functionally segmented with little integration of resources or joint planning.[41] Indeed, the Chief of the Defence Staff (CDS), General Sir Nick Carter, eloquently surmised the nature of warfare and logistics at this time:

> When I grew up in the Cold War, it was straightforward. We were at four hours' notice to move, we sat in our barracks in Germany, we knew where all our equipment was, we knew where our deployment positions were and we were ready to go for a very clear and present threat that we understood.[42]

---

40   Moore, D. and Antill, P. (2000) 'Where do we go from here? Past, present and future logistics of the British Army', *The British Army Review*, 125, 67.

41   Ibid.

42   *The Daily Telegraph*, 19 January 2016, 'Legal action against soldiers "could undermine Britain on the battlefield" warns chief of general staff'.

Figure 3.1   NATO 'layer cake' defensive plan
Source: U.S. Office of Technology Assessment, 1987.

In short, notwithstanding differences in the availability of strategic airlift and force posture, the logistics system of NATO in the Cold War was similar to that of the Allies during the European campaign of 1944–45 in that it was predominantly reliant on depots, trucks, and the segmented functions approach.

Despite the end of the Cold War and the introduction of some early post-Fordist practices, Western military logistics in the 1991 Gulf War still remained organised around the Cold War echelon system. Although the campaign was expeditionary in nature, the coalition still could deploy and build up its forces in secure areas away from the frontline in Kuwait. These forces were likewise supplied by secure logistics bases and lines of communication, and combat operations were directed against a linearly deployed conventional enemy (the Iraqi army) using defensive tactics to hold national territory. Due to this operational reality, the US logistics system operated out of the Saudi port of Jubail, through Al-Qaysumah base, and then moved goods onto divisional logistics bases. To keep the 700,000 US troops supplied, 18 trucks per minute, 24 hours a day, seven days a week passed on the main supply route.[43] As such, at first glance, it appears the Cold War model still applied.

---

43   Pagonis, *Moving Mountains*, 9.

Nevertheless, the overarching impression of the US logistics operation given by the officer responsible, Lieutenant General William Pagonis, is of a logistics system innovating and adapting under the pressure of sustaining such large forces in the desert. Contrasting the focus on the RMA combat technologies, Pagonis notes that the lack of asset tracking systems resulted in massive unused stockpiles, while he states that the whole logistics plan and detailed schedule was still recorded in paper format in a single 'red book' binder.[44] Pagonis also tells how he had to develop logistics planning cells during deployment to assess logistics requirements, analyse activities and draw up contingency plans, indicating that these cells were an innovation rather than determined by logistics doctrine. However, Pagonis' account does indicate the emergence of some post-Fordist thought in Western military logistics at this time. He refers to combat soldiers as his 'customers', and also describes how the up-arming of the Abrams tank with the new 120mm gun was achieved by the movement of the entire production line from the US to Saudi Arabia and the adoption of round-the-clock production, indicating a growth in contractor flexibility.[45] Similarly, the massive logistical needs of the U.S. Army forced Pagonis to meet demand through local sources where possible. He details how vehicles were mass-rented from Saudi firms by open market negotiation, noting how the Saudis often took advantage of the U.S. Army's demand to manipulate prices.[46]

The British experience in the Gulf War paints a picture of a military adapting Cold War logistics doctrine to a new environment in a similar fashion to that of the US. The lasting impression given by the commander of British forces, General Rupert Smith, is of a logistics plan that struggled to maintain and supply an armoured division in the Gulf and one that may have not survived contact with a more competent enemy.[47] The British Army's Gulf logistics plan followed Cold War doctrine in its adoption of echeloned rear bases and three lines of supply to support a linear battle.[48] A single theatre supply area, known as the Force Maintenance Area (FMA) was initially established at Jubail, and supplies were to be trucked to a Forward Force Maintenance Area (FFMA), and then to a Divisional Maintenance Area for distribution to frontline units along secure lines of communication. Each echelon had their own contingent of engineer, transport, logistics and medical units, and the FMA also had a team dedicated to procuring local supplies and services. Just In Case logistics was still practised, with the FFMA stockpiled with enough supplies to sustain the division in combat for at least 10 days.[49]

---

44 Ibid., 104–106.
45 Ibid., 8–9.
46 Ibid., 106–108.
47 Smith, 'The Commander's Role', 23.
48 Harber, Colonel W. (1995) 'The Logistic Structure' in White, *Gulf Logistics*, 34.
49 White, Major General M. (1995) 'The Support of the War in Perspective' in White, *Gulf Logistics*, 9.

Figure 3.2   British Army logistics convoy during the Gulf War
Source: U.S. Federal Government Copyright. 1991. PHC Holmes.

However, the UK logistics effort also suffered from poor inventory and asset tracking procedures, with stores held in depots in Germany since the end of the Second World War frequently missing or unserviceable.[50] Some weapons platforms, like the new Challenger tank, required desert upgrades in theatre; a task that was completed with the deployment of specialist teams from British firms.[51] Meanwhile, the mechanical spares system was 'overwhelmed' due to a lack of asset tracking systems.[52] Asset tracking was a major flaw across British logistics, with a single officer forced to examine each container at Jubail to find critical medical supplies.[53] Movement control IT systems were also incompatible: in the words of one British logistics officer, this resulted in manual information gathering 'using stubby pencil, T cards and the most famous ... computer of all, fagpacket [becoming] the day to day tool of the mover.'[54] Smith highlights how many ships

---

50 Smith, 'The Commander's Role', 20.
51 White, 'The Support of the War in Perspective', 25; Campbell, Colonel A. (1995) 'Equipment Support' in White, *Gulf Logistics*, 148.
52 Campbell, 'Equipment Support', 151.
53 Lillywhite, Colonel L. and Leitch, Colonel R. (1995) 'Medical Support' in White, *Gulf Logistics*, 94.
54 Reehal, Colonel P. (1995) 'Transport and Movement' in White, *Gulf Logistics*, 67.

were loaded to capacity to reduce costs rather than in the order their stores would be needed for operations, thus hampering tactical flexibility. Nevertheless, there is also evidence that British logistics was evolving during this period. Smith personally oversaw the development of an armoured logistics reserve battlegroup to negate the dangers of operating on insecure lines of communication once the offensive had begun, a scenario that contradicted British logistics doctrine at the time.[55] Meanwhile, new transport technologies, such as the Dismountable Off-Load and Pick-Up System logistics truck also significantly increased the speed and flexibility of British logistics operations.

Although there were some signs of innovation and adaptation, the Gulf War highlighted that Western military logistics were still fundamentally based on the echelon system of re-supply in secure rear areas and JIC practices. The same system was followed in the US and UK deployments on NATO's Balkan missions, and the British deployment to Sierra Leone. Indeed, it appears that the gap between industrial and military logistics during this period was at its widest, and despite attempts to centralise logistics command, exemplified in the development of the joint force U.S. Transportation Command (USTRANSCOM), functional segmentation remained high. US and British military logistics organisations had been relatively insulated due to large budgets and the need to counter a single, constant existential threat in the Cold War. But by 1999, with shrinking budgets and the desire for a smaller, more globally deployable military to address increasingly diverse threats, it became apparent that their logistics organisation and systems were far behind the curve. In February 1999, the UK began centralising logistics in its new Defence Logistics Organisation, whose main tasks were to streamline defence logistics structure, reduce stock costs and manage procurement reform.[56] In the US that same month, a military commander called for a distribution-based 'Revolution in Military Logistics'.[57] Heeding this call, in 2000 the U.S. Department of Defense launched a 'Logistics Transformation Plan' in order to modernise logistics.[58]

Despite these early attempts to reform military logistics, the 2003 Iraq War highlighted major shortcomings in both nations' systems. These failures became embedded in public perceptions, with headlines such as 'Families of dead soldiers can sue MoD over inadequate kit'; 'US soldiers lack best protective gear' and 'Thousands of Army Humvees Lack Armor Upgrade' highlighting media interest.[59]

---

55 Smith, 'The Commander's Role', 20.

56 House of Commons Defence Select Committee (2000) *Second Report: Ministry of Defence Annual Reporting Cycle*, London: HMSO, paras 135–38.

57 O'Konski, M. (1999) 'Revolution in Military Logistics: An Overview', *Army Logistician*, Jan.–Feb. 1999.

58 U.S. Department of Defense (2000) *Defense Reform Initiative Directive No. 54 – Logistics Transformation Plans*, Introduction, 2.

59 *USA Today*, 17 December 2003, 'US soldiers lack best protective gear'; *The Times*, 20 October 2012, 'Families of dead soldiers can sue MoD over inadequate kit'; *The Washington Post*, 12

Reacting to public concern, a House of Commons inquiry into British preparations for the invasion of Iraq noted that as: 'a result of a combination of shortages of initial stockholdings and serious weaknesses in logistics systems troops at the frontline did not receive sufficient supplies in a range of important equipment including enhanced combat body armour ...'[60] A British Commanding Officer during the initial war-fighting phase went further, describing the delivery of logistic support to frontline operations as 'woefully inadequate'.[61] The impact of the failure of unresponsive 'brute force' logistics based on JIC principles was not only felt by British troops. A U.S. Congress investigation found that in the first month of combat operations, the defence department temporarily 'lost track of $1.2 billion in materials shipped to the Army, encountered hundreds of backlogged shipments, and ran up millions of dollars in fees to lease or replace storage containers because of backlogged or lost shipments.'[62] Other inefficiencies identified included port congestion, improper sequencing of combat units and their support, excess costs and the disrupted flow of units and supplies into theatre.[63] Clearly, Western military logistics were failing, and the perception was they were failing because they had not adopted industry best practice. As the conflict in Iraq continued, the need for more cost-efficient logistics became increasingly important. A real RML was needed, and both the American and British militaries began to transform their logistics systems to emulate post-Fordist industry.

Centralisation

The presence of neo-liberal governments in both the US and UK who were committed to outsourcing state functions to private industry is important in understanding the wider drive for military logistics efficiencies. Similarly, the RML could not have taken place without the vast increase in computing and information technology capacity. But, more specifically, the media coverage of early logistical failures in the invasion of Iraq and the resulting political pressure to address the issue ahead of presidential elections, forced US Secretary of Defense, Donald Rumsfeld, to prioritise logistics transformation. In September 2003 Rumsfeld began the process of centralisation, designating Commander, USTRANSCOM as responsible for all distribution across US defence. In defining USTRANSCOM

February 2007, 'Thousands of Army Humvees Lack Armor Upgrade'; *The Guardian*, 26 July 2009, 'Lack of helicopters puts injured troops at risk'.

60 Ministry of Defence (2004) *Operation TELIC — United Kingdom military operations in Iraq*, London: HMSO, 4.

61 Yoho, K., Rietjens, S. and Tatham, P. (2013) 'Defence logistics: an important research field in need of researchers', *International Journal of Physical Distribution and Logistics Management*, 43(2), 85.

62 *Federal Times* (9 May 2005) 'DoD Told to Shape Up'.

63 Simon, 'The Art of Military Logistics', 64.

as the Distribution Process Owner (DPO), Rumsfeld ensured it became 'the single entity to direct and supervise execution of the Strategic Distribution system' in order to 'improve the overall efficiency and interoperability of distribution related activities – deployment, sustainment and redeployment support during peace and war.'[64] Rumsfeld had experience of transforming ailing businesses by streamlining procedures and reducing workforces in line with post-Fordism, and his initiative was informed by his awareness that the entire US defence distribution pipeline needed to be properly linked and synchronised to produce the most cost-effective means of supply. In short, it had to emulate industrial logistics by incorporating new information technologies into newly centralised organisations and systems.

To this end, the reforms designated USTRANSCOM's four-star general as the single, unified commander for all defence distribution, and outlined a four-year plan to change organisational structures and upgrade IT systems to give complete oversight of the distribution system. Paradoxically, enabled by centralised and standardised IT systems, the decentralisation of decisions throughout the distribution pipeline encouraged the logistical flexibility to respond quickly to frontline demands.[65] The centralisation and standardisation of logistics under USTRANSCOM continued in 2004 when the organisation became the manager of all US defence logistics information technology systems. In 2006 it was made responsible for identifying, recommending and supervising implementation of all global sourcing solutions. Decisively, at this time USTRANSCOM adopted the civilian Supply Chain Operations Reference Model (SCORM) which identifies core institutional processes and tailors production-supply chains to meet these processes.[66]

The British military also increased the pace and scope of centralisation in response to the shortcomings of Iraq. In 2004, the Defence Logistics Transformation Programme (DLTP) was launched with the aim of increasing the effectiveness, efficiency and flexibility of logistics support across UK defence, appointing a single joint four-star officer, the first Chief of Defence Materiel. The DLTP began the process of centralising defence materiel and resources and created centralised centres for the repair and maintenance of weapons platforms. While the DLTP did increase effectiveness with the introduction of the centralised JAMES whole fleet management and the VITAL asset tracking systems, it was primarily centred on cost-reducing efficiencies, and, supported by consultants at McKinsey, it eventually delivered savings of £952 million.[67] A renewed focus on effectiveness

---

64 Available at http://ustranscom.mil. retrieved 20 November 2014.

65 Smith, B. (2007) 'The Mandate to Revolutionize Military Logistics', *Air and Space Power Journal*, 21(2), 92.

66 Maddox, E. (2005) 'Organizing Defense Logistics: What Strategic Structures Should Exist for the Defense Supply Chain', unpublished thesis, 5.

67 McKinsey and Company (2010) 'Supply Chain Transformation Under Fire', available at https://www.mckinsey.com/~/media/alumni%20center/pdf/mog_supply_chain.ashx retrieved 29 November 2014.

came with the introduction of the Defence Logistics Programme in 2006. This sought to increase coherence, velocity and precision across logistics through the centralisation of command and control, and the updating and centralisation of IT systems.[68] Emulating the US, 2007 saw the merger of Britain's two defence logistics organisations into a single entity, Defence Equipment & Support. Responsibility for operational logistics was also centralised in the Permanent Joint Headquarters J4 division and within the theatre-deployed Joint Force Logistic Component Headquarters.[69] Meanwhile, a single centralised inventory system for the whole of UK defence, the Management of Joint Deployed Inventory (MJDI), was commissioned to provide one platform to link previously incompatible asset tracking systems. MJDI aims at total asset visibility to enable British defence to move to a fully JIT logistics system. It will lead to a profound re-organisation and flattening of hierarchies in British logistics units, and also a reduction in combat units' logistics personnel, thereby delivering efficiencies. Similarly, NATO logistics command has also begun centralising, although at a much slower pace. In 2011, member states agreed to reduce and re-organise the alliance's logistics structure into the NATO Procurement and Support Agency (NSPA). This combined four former NATO logistics commands into one. Clearly, then, the US, UK and NATO have centralised their logistics commands, organisation and systems in line with Fordist principles.

Another major example of ongoing centralisation in British defence particularly pertinent to this study has been the recent implementation of the 'Defence Estate Rationalisation' programme. The 2010 SDSR identified the need for rationalisation of property owned by the MoD in order to save running costs, and primarily recommended the sale of surplus land and buildings that could be undertaken quickly. To achieve this, the Army Basing Plan was announced in March 2013.[70] This plan influenced FR20, which outlined the centralisation of Army Reserve units in larger barracks and the closure and subsequent sale of smaller sites. Of the 334 TA sites around Britain in 2013, FR20 designated 26 to close, most of which were done so on the grounds that the units that occupied these locations were 'under-recruited.'[71] As is discussed in Chapter 6, this sale of some of the defence estate, coupled with the centralisation of equipment stores in larger bases under the Whole Fleet Management approach, has had major impacts on some of the sub-units in this study. Supporting the argument made in the previous chapter, and the evidence

---

68 Leeson, Air Vice Marshal K. 'Multi-National Logistics Transformation – A United Kingdom Update', Briefing available at https://ndiastorage.blob.core.usgovcloudapi.net/ndia/2006/logistics/leeson.pdf, retrieved 4 March 2016.

69 Ministry of Defence (2012) *Joint Service Publication 886: Defence logistics support chain manual*, London: MoD, 6.

70 Ministry of Defence (2013) *Defence Estate Rationalisation Update*, available at https://www.gov.uk/government/publications/defence-estate-rationalisation-update, retrieved 28 July 2016.

71 *Hansard*, 3 July 2013, col. 924.

provided in the next chapter, this decision was also politico-ideological, reflecting as it did the Conservative government's desire to reduce state spending overall. Indeed, this became especially obvious after they gained a majority in the 2015 general election, and quickly introduced a second round of rationalisation in March 2016. This included the potential sale of ten sites facilitated by the relocation of regular units from their barracks into centralised super-garrisons. This marked a major departure from simply selling under-used sites or relocating reserve units, as the 'release' of these sites is expected to generate £1 billion through land sales. It was also designed to complement the Conservative government's house building scheme by contributing 'up to 55,000 homes to support wider Government targets' by 2020.[72] Clearly then, the British government has adopted defence centralisation in earnest.

Integrating the Core and the Periphery

As with the combat arms, a core, specialised logistics workforce is being established in Western militaries. Enabled by better training and technology, these core logistics organisations are professionalising the study and practice of logistics. Highlighting a major shift in institutional goals, this new core is increasingly specialising in the management of logistics IT systems and contracting. For example, the U.S. Army's Materiel Command is expanding the training of its cadre of in-service contracting professionals to increase the capability of the Army to understand and engage with its contractors.[73] The establishment of a Logistics Contract Management Course run by the Defence Logistics School indicates that British forces are following suit. In a further sign of specialisation, the U.S. Army has consolidated previous logistics learning environments by opening its own logistics university in 2009, which runs over 200 specialised logistics courses. Similarly, the introduction of MJDI in the British Army will be enabled by the creation of a specialist unit to support higher command. The full roll-out of MJDI will also change logistics structure at the unit level, with new Logistics Support Detachments (LSDs) embedded with each unit. These detachments will consist of a team of four 'professional logisticians' and will replace the old system of each unit providing their own non-specialist logistics staff.[74] This smaller, core LSD will significantly reduce the number of logistics-related personnel in each unit, and will deploy with its parent unit, resulting in significant changes to the way army units are supported. Meanwhile, the UK, US and NATO have also been careful to maintain core logistics functions deemed central to operational effectiveness. The focus on the core, then, appears to have utility.

---

72 *Hansard* (2016) 'Defence Estate Rationalisation', available at https://www.parliament. uk/business/publications/written-questions-answers-statements/written-statement/ Commons/2016-03-24/HCWS659/, retrieved 28 July 2016.

73 Dunwoody, 'Strategic Choices', 84.

74 British Army, *Tactical Logistics Support Handbook*, para. 70.

Simultaneously, due to defence budget cuts, the US and Britain have had to sharply reduce their logistics forces' size while attempting to maintain their capability. These cuts have often focused on the logistics component precisely because it is perceived that much of the non-core capability can be provided by a periphery workforce.[75] To reduce costs, many expensive and traditional logistics functions needed during large mobilisations have been allocated to reserve forces, which have been increased in size in a bid to maintain capability. For example, while the Army2020 reforms reduced the British Army's logistics personnel by about 30 per cent, the complementary reform of the Army Reserve has increased reliance on reserve logistics units to meet surges in demand, and led to the creation of many new reserve logistics units.[76] Indeed, reservists now constitute six of the 13 REME battalions.[77] While the UK's reserve logistics component has therefore been increased (but not necessarily filled), the capabilities it provides have generally remained toward the lower end of the skill spectrum. Those units that do have a more specialised function have been formed with the specific aim of incorporating previous military or civilian skills into current capability to reduce costs. As such, the delegation of lower skilled logistics functions to reservists, coupled with the desire to tap the specialised ex-military or civilian workforce, indicates the dual nature of modern military logistics structural reliance on the periphery.

However, both the centralisation of command and the division of labour between the core and the periphery have been complemented by the adoption of the 'total cost' approach to force structure and readiness in Western militaries. The closer integration of the core and the periphery is underpinned by the total cost approach, and supported by SCM practices that strive for the flexibility to respond to consumer demands while keeping running costs down. British defence is currently being reorganised around the 'Whole Force' concept that became fully operational in April 2014. This is focused on ensuring that the product – which in the military sense is manpower – consists of 'the right mix' of 'regulars, reserves and contractors to produce the greatest effect in the most cost-effective manner.'[78] For example, the total British deployed force on Operation Herrick in Afghanistan between 2010–2014 consisted of almost 78 per cent regulars, nearly 20 per cent reservists, and less than 2.5 per cent contractors and civil servants.[79] By achieving a more 'balanced mix' in the future – itself a term that appears to have originated in

---

75 For example, Future Force 2020 cut the RLC personnel by 25 per cent and engineers by 30 per cent. *The Daily Telegraph*, 31 March 2012, 'Britain's most famous regiments spared in defence cuts'.

76 British Army, *Tactical Logistics Support Handbook*, para 80; *Future Reserves* 2020 para 2.32.

77 British Army (2013) *Battlefield Equipment Support Doctrine*, 22.

78 Parkinson, Brigadier R. (2012) 'Affordability in the International Environment', available at http://www.dtic.mil/ndia/2012logistics/Parkinson.pdf, accessed 30 November 2014; Ministry of Defence (2014) 'How Defence Works', Version 4.1, page 8.

79 Briefing by Director Medical Services, available at https://www.rusi.org/downloads/assets/ Jarivis_part_2_FINAL_use_this_one.pdf, accessed 30 November 2014.

the energy industry – UK defence planners hope to retain capability and flexibility while decreasing expensive manpower costs. The Whole Force concept is enabled by operational planning assumptions, tiered levels of force readiness, and better trained and equipped reserve forces. The logistics element, the new Total Support Force (TSF), follows the same approach. The TSF comprises a 'pre-planned mix of military, civil service and contractor personnel held at appropriate readiness to provide progressive levels of support in the UK and on operations.'[80]

In 2012, the U.S. Army also instigated the 'Total Force' policy aimed at better integrating the Army, National Guard and Army Reserve components below the divisional level. Specifically, these reforms standardise reserve readiness with those of the army, and place responsibility for validating this readiness with army command.[81] Under analogous fiscal pressures, and following advice from business management firm Price Waterhouse Coopers, the Irish Defence Forces also introduced a 'Single Force Concept' in 2012.[82] German, French, and NATO's defence structures have not yet adopted the concept, predominantly due to ongoing reserve transformation in the case of the former and political issues in the latter two, but the recent adoption of the 'total cost' approach by some Western militaries represents a profound change in the way in which not only their logistics, but also their wider military forces, are organised, resourced and deployed. In the US and UK, the change has occurred in a similar time frame due to similar budgetary pressures and strategic appraisals. While the total force approach therefore has implications for military logistics, the force structure solutions they provide are based on a post-Fordist logistics organisation. Coupled with the centralisation of logistics commands and the division between core and periphery logistics functions, the adoption of the integrative 'total cost' approach in Western militaries signals the end of the segmented function logistics of Van Creveld's era.

SCM and Outsourcing

The success of the total cost approach relies on two decisive criteria being met. Firstly, there must be a comprehensive understanding of demand, and secondly there must be an understanding of how this demand will be met. Forecasting and supply are thus crucial. At the strategic level, Western military logistical demands are set by force structure and strategic appraisals, such as the SDSR and the Quadrennial Defense Review. At the operational level, logistics forecasting is demand-based. This has changed little since the Second World War. However, the

---

80   British Army, *Tactical Logistics Support Handbook*, para. 39.
81   U.S. Secretary of the Army (2012), *Army Directive 2012–08 Total Force Policy*.
82   Irish Department of Defence (2012) *A Value for Money Review of the Reserve Defence Forces*.

supply side has changed dramatically with the introduction of SCM principles and systems into military logistics. Indeed, without the transparency and oversight of supply encouraged by SCM, the integrated total force concept would be impossible to implement.

British logistics practices are now heavily dependent on SCM systems. After the 2003 Iraq deployment, McKinsey was heavily involved in introducing SCM procedures across British defence, with a particular focus on increasing delivery reliability while decreasing wait time. Numerous procedural inefficiencies were identified, along with the need to update IT systems. One of the most noticeable changes under the SCM approach occurred in relation to unit stores. Under the previous segmented approach, units held 30 days of stores in contingency. However, by linking existing demand data with engineering analysis and the experience of quartermasters, standard stores and bespoke 'priming equipment packs' are now kept within the supply chain, giving far more flexibility. Reflecting the desire to move to a wider SCM footing, in 2005 the Joint Supply Chain concept was introduced by the Ministry of Defence to 'cover the policies, end-to-end processes and activities associated with receipt of stocks from trade to their delivery to the demanding unit and the return loop for all 3 Services.'[83]

Similar changes have been underway in the US. In May 2003, the DoD published its Supply Chain Materiel Management Regulation outlining the conduct of future joint logistics. This introduced SCORM, and at its core was an awareness that US defence logistics needed to be more responsive, reliable and consistent to adapt to the evolving global environment while delivering the best value for money. In 2007, the first of three phases in the introduction of SCM, the Joint Supply Chain Architecture was initiated, and in 2010 it was institutionalised. The most recent DoD manual on SCM procedures instructs the military to 'monitor and adopt or adapt emerging business practices to provide best-value, secure materiel and services, improve DoD supply chain performance, and reduce total life-cycle systems cost.'[84] Thus the US military clearly aims to continue to emulate industry to increase efficiency. While NATO is yet to adopt a total force structure or a full SCM approach, its logistics principles indicate the impact of SCM concepts on its doctrine.[85]

Coupled with the outsourcing of logistics capability to reservists, the British Army's new TSF puts a similar emphasis on contractors. Crucially, the new structure states that: 'the use of non-military personnel will [provide] most if not all logistics functions rear of the Theatre Support Group by roule 4 of an enduring operation.'[86] Therefore, the British Army's doctrine is to delegate rear

---

83   Ministry of Defence, *Joint Service Publication 886*, 3.
84   U.S. Department of Defense (2014) *DoD Supply Chain Materiel Management Procedures: Operational Requirements*, p. 7 enclosure 2.
85   NATO *Logistics Handbook 2012*, 50–51.
86   British Army, *Tactical Logistics Support Handbook*, para. 45.

logistics functions to the private sector by the second year of a deployment. This doctrine also states that reliance on Contractor Support to Operations will increase in ratio as the army decreases in size due to recent cuts. Decisively, it states that early engagement with long-term contractors during operational planning is required, and that contractors should be included in the whole spectrum of these plans, from force generation, to deployment, sustainment and force protection. That outsourced contractors are now involved in military planning is a potentially significant change in the relationship between industry and the military. Meanwhile, there has also been a transformation in the nature of outsourced contracts. With the support costs of complex weapons systems now exceeding the cost of development and production by two to three times over their service life, Andreas Glas et al. have shown that Performance-Based Logistics (PBL) contracts are becoming increasingly common in Western militaries as a method of reducing these costs. For example, PBL can mean that a civilian firm is contracted to deliver a required amount of flying hours on an airframe, rather than hours of servicing.[87] Similarly, in their analysis of NATO's outsourced contracts in Afghanistan, Christiaan Davids et al. show that while member states often conducted independent sourcing, pooled operational sourcing was also commonly used.[88] The pooling, sharing and prior negotiation of outsourced logistical services in the US, UK and NATO indicates the adaptation and increasingly privatised nature of military logistics due to constrained fiscal realities.

While others have discussed the new reliance on contractors in detail, it is important to note here that the scale of contractor support to both the US and British militaries in Iraq and Afghanistan was politically useful as it meant troop caps could be circumvented in these unpopular wars.[89] Meanwhile, the fact that the second and third largest winners of US contracts in Iraq were both Kuwaiti firms, and that contracting in Afghanistan had to be reorganised due to corruption allegations, highlights how the awarding of contracts to local suppliers can influence reconstruction efforts. In modern conflicts, military logistics – previously undertaken by military forces – can be incorporated into strategic goals through their generation of economic activity.[90] No longer confined to sourcing nationally, Western militaries have demonstrated their ability to leverage the globalised economy through outsourcing. This has potentially major implications for Western strategy.

---

87 Glas, A., Hofmann, E. and Essig, M. (2013) 'Performance-based logistics: a portfolio for contracting military supply', *International Journal of Physical Distribution & Logistics Management*, 43(2).

88 Davids, C., Beeres, R. and van Fenema, P. (2013) 'Operational defense sourcing: organizing military logistics in Afghanistan', *International Journal of Physical Distribution & Logistics Management*, 43(2), 125.

89 Cusumano, 'Bridging the Gap', 111.

90 Erbel and Kinsey, 'Think Again', 21.

The Emerging Logistics Network

Very recently, the focus on the supply chain has been replaced by the realisation that more networked logistics will be a crucially important enabler in future conflict. Indeed, even before the introduction of JIT and SCM, US commanders were aware that the ultimate goal of these processes was a 'seamless logistics system that ties all parts of the logistics community into one network of shared situational awareness and unified action.'[91] As its' latest logistics doctrine, Joint Defence Publication 4.0 (JDP 4.0) indicates, the British military is now taking steps to move beyond SCM by creating a fully networked logistics system which encompasses more than just the supply chain. Highlighting this, according to one senior officer responsible for transforming British Army logistics, 'networking is the new buzzword'.[92] As JDP 4.0 states, 'Logistics stretches across a network of nodes with multiple processes, through which personnel and materiel flow and services are provided'.[93] Thus, the whole British military and accompanying international logistics system is now conceptualised as a 'support network' of interconnected nodes of suppliers, consumers, maintainers and storers. This network approach seeks to eclipse SCM by moving beyond the supply chain to a more expansive view of all supporting producers, services, and partners while simultaneously allowing supplies to be moved 'forwards, backwards and sideways' between nodes. Rather than only moving supplies forwards toward the end user: 'the network spreads the load' associated with potentially stove-piped supply chains by allowing storage within its nodes, thereby reducing logistic drag.[94] Contrasting the logistics overlay of the NATO's 'layer cake', Figure 3.3 shows that in a globalised world, it is recognised that the British military supply network must be global. However, this network-enabled capability must be supported by ever more complex IT systems with open architecture across nodes – a situation yet to be reached. To be effective, a fully networked system also needs accurate consumption and environmental data which is largely missing at present. Thus, the British Defence Support Network is still under development; it is not yet a fully networked strategic supply system.

Nevertheless, the introduction of the MJDI and Total Asset Visibility Minus (TAV-) systems across British defence is another good example of an emerging logistics network. MJDI will replace the stockpiling and stove-piping associated with segmented function logistics across all units and formations, leading to a truly networked logistics IT system. It will allow for the total global visibility of all stock up to unit level, and therefore better asset management. It will also be

---

91 Reimer, J. (1999) 'A Note from the Chief of Staff of the Army on the Revolution in Military Logistics', *Army Logistician*, Jan./Feb.
92 Interview, senior British Army logistics officers responsible for future doctrine, Andover, 9 June 2015.
93 Ministry of Defence, *Logistics for Joint Operations*, 9.
94 Interview, senior logistics officers, 9 June 2015.

inter-operable with TAV- which uses tagged barcodes on vehicles, containers and pallets that can be read by radio frequency. This allows the automatic logging of all supplies as they pass through TAV- nodes, in stark contrast to the experience of the Gulf and Iraq Wars. This visibility, linked with the MJDI system, will allow demand to be judged in near real-time, and allow logistics planners to move stocks from one unit to another based on priorities and requirement data within the system, rather than solely on the demands of the units.[95] Crucially, by 'turning every unit into a secondary depot', MJDI aims to create a distribution network across British defence, with every node in the network able to see what is in the system and where it is at any time. Compared to the segmented functions system, MJDI has the potential to completely transform both the structure and procedures of British defence logistics, bringing them in line with industry best practice. While space precludes a detailed discussion of the operational level manifestation of the network approach to logistics here, it is simply important to note that contrasting the single theatre entry points of the past, where possible and as in Afghanistan where the Northern Distribution Network was adopted to supply NATO, Western nations now prefer multiple entry points to increase both logistical flexibility and resilience to geopolitical issues.[96]

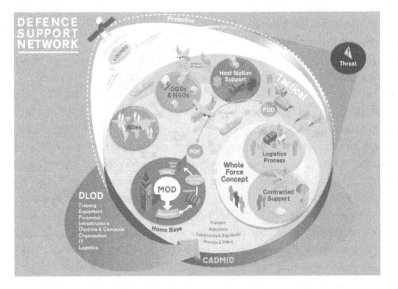

Figure 3.3   The British Defence Support Network

Source: MoD Crown Copyright 2015. Defence Science and Technology Laboratory.

---

95 British Army, *Tactical Logistics Support Handbook*, paras 65–69.
96 Center for Strategic and International Studies (2009) Report, *The Northern Distribution Network and the Modern Silk Road: Planning for Afghanistan's Future*, Washington, DC: CSIS.

Recent Tactical and Operational Logistics

While it is clear that a major change in the way military logistics organisation and systems are structured and managed at the strategic and operational levels has occurred, there is also evidence to suggest that tactical logistics practices have adapted to the modern battlefield. As the conflicts in Iraq and Afghanistan were defined by insurgencies operating in non-linear battle spaces, Western forces relied on Forward Operating Bases (FOBs). Although the FOB concept itself dates back to the British Army's isolated imperial outposts, when faced with unconventional insurgent tactics emphasising surprise attacks from within the population, Western militaries resorted to static, forward, defendable bases in recent expeditionary conflicts. Given the lack of clear frontlines between opposing forces, the FOB system relied on supply from theatre bases, such as Shaibah Logistics Base in Iraq, or Camp Bastion in Afghanistan, which were heavily defended and preferably sited far from the population. However, this meant that – unlike in the rear areas of linear battlespaces – these bases could not take advantage of existing infrastructure and were often situated far from the FOBs. As a result, these bases had to create their own water, energy and fuel infrastructures, and although supplying smaller forces, the logistical burden was therefore greater. Indeed, one British logistics officer has described the task of constructing Camp Bastion, the largest logistical base since the Second World War, as similar to building 'Aldershot with Gatwick [airport] bolted on ... in the face of a lethal insurgency in a landlocked country.'[97]

Due to similar tactical pressures, beginning in Iraq, and most notably in Afghanistan, the system of FOBs and smaller COBs (combat outposts) significantly altered both combat and logistics operations. Erbel and Kinsey noted the logistics problems in the Helmand campaign, but did not discuss the FOB supply system in detail. Robert Egnell and King both examined how the British 'ink-dot' operational plan relied on FOBs to disperse limited forces across large, unsecured areas of Helmand province.[98] While these provided a defendable and secure base for troops in often highly volatile areas, most of the FOBs were unable to mutually support one another. Coupled with this, vast areas, including main supply routes, could not be adequately secured with the small forces involved. Leonard Wong and Stephan Gerras have argued that with the insurgent threat seemingly everywhere and nowhere, the safety of the FOB, with its better food, chance of rest and its provision of technologies to communicate with families,

97 *The Sunday Times News Review* (2 November 2014) 'We Think It's All Over'.
98 Egnell, R. (2011) 'Lessons from Helmand, Afghanistan: What now for British Counterinsurgency', *International Affairs*, 87(2); King, A. (2010) 'Understanding the Helmand campaign: British military operations in Afghanistan', *International Affairs*, 86(2).

has changed Western soldiers' experiences of war.[99] Yet precisely because of the supplies needed to feed and defend a FOB, and the services expected in them, the FOB has created its own logistical challenges.

Most obviously, the FOB system resulted in the flattening of the conventional hierarchical echelon logistics system with secure supply lines, which was largely replaced by a more nodal method of re-supply. For example, while a company in a FOB would usually first request re-supply from its own chain of command, the fact that almost all commands and their respective stores were usually co-located in the main theatre base, rather than at different points in an echelon, meant that it was usually quicker and easier to receive supplies compared to the old echelon system. This centralisation of logistics command and stores in theatre was also accompanied by a centralisation of the means of delivery. Despite Van Creveld's assertion that helicopters were unlikely to affect logistics due to their cost, support helicopters were frequently pooled between different nations and commands at bases to provide greater re-supply capacity and flexibility. Re-supply road vehicles were also pooled. Instead of supplies travelling up one unit's echelon system, often a single logistics convoy dropped supplies off at a series of different units in different FOBs before returning to the main logistics base. Although the basic principles remained the same, in Afghanistan nodal distributed logistics replaced the linear, echeloned supply system.[100] This was driven in response to the constraints of terrain and insecure lines of communication, and the changing post-Fordist organisation and systems of the RML.

While the use of support helicopters to re-supply FOBs is an obvious method of negating insurgent attacks on logistics convoys, it remains an expensive means of logistics delivery. As a result, traditionally lightly-armed logistics convoys have been adopted to vastly increase their firepower and armour. This has led to what could be termed the 'combatification' of tactical logistics in modern conflicts, highlighted by the increasing use of Combat Logistics Patrols (CLPs). Known as 'clips', CLPs have become enshrined in British logistics doctrine as the favoured method of operating in insecure areas.[101] These convoys can stretch miles long and often consist of over 200 heavy vehicles travelling distances of up to 200 kilometres through unsecured territory. Vehicles usually consist of up-armoured military trucks like the new, crane-equipped British MAN Supply Vehicles, accompanied by other specialist logistics vehicles such as the OSHKOSH series of transporters. For force protection, these vehicles are now accompanied by the heavily armed and armoured MASTIFF or MRAP protected mobility vehicles, containing fighting troops. At the head of the convoy, an

---

99   Wong, L. and Gerras, S. (2006) *CU @ The FOB: How the Forward Operating Base is Changing the Life of Combat Soldiers*, Washington, DC: Strategic Studies Institute, 25.

100   Interview, senior British Army logistics officers responsible for future doctrine, Andover, 9 June 2015.

101   British Army, *Tactical Logistics Handbook*.

anti-Improvised Explosive Device (IED) roller vehicle can be used to detonate pressure plate devices. Tanks and AIFVs can also accompany the convoys. As Figure 3.4 shows, upscaling for combat has not only clearly transformed supply operations, it has affected the training and posture of logistics soldiers who now may need to fight their way through to FOBs. In Afghanistan, the tactical threat forced Western militaries to quickly procure these vehicles in large numbers and at considerable cost, but it also altered the nature of combat operations in the FOBs themselves. With CLPs occurring every week to keep FOBs supplied, the few available combat troops were frequently detailed to secure the main supply routes for hours each time the convoys passed.[102] Such was the money, time, and lives expended on resupplying the FOBs that, to paraphrase Clausewitz, in many respects 'it was as if the whole [British] war-engine had ventured into the enemy's territory in order to wage a defensive war for its own existence.'[103] This is on such a scale that the tail can wag the dog, with profound implications for military effectiveness at the tactical, operational and strategic levels.

Figure 3.4 A British Combat Logistics Patrol in Afghanistan
Source: Sgt Wes Calder RLC, MoD Crown Copyright 2012.

Strategic Vulnerability?

This chapter has shown how the current dominant mode of production, post-Fordism, is a useful conceptual framework for understanding the organisational blueprint and systems that enable the RML. In doing so, it is clear that in

---

102 Bury, P. (2010) *Callsign Hades*, London: Simon & Schuster, 237–39.
103 Van Creveld, *Supplying War*, 28.

emulating industry practice, the US, UK and NATO have transformed their logistics organisation, systems and tactics since the Gulf War. Military logistics has been commercialised and civilianised through the processes of centralisation; integrating the core and the periphery; outsourcing and SCM; and the emergent logistics network. Enabled by advances in technology, these processes continue to occur, often at different paces and to different extents across these militaries, but all following broadly similar goals and trajectories. The cumulative effect of each of these processes has resulted in the most profound change in military logistics since the introduction of the combustion engine. Furthermore, with the adoption of the whole system approach to both military logistics and wider military forces, it is clear that the potential impact of each process is heavily dependent on the introduction of the others; for example, centralisation enables networks, networking can enable outsourcing capacity. As a result, the whole systems approach is creating a logistics system, but also a wider force structure in the West that relies on high levels of integration to generate the most efficient capability from smaller organisations within tighter time frames. This transformation has important implications for the military logistics literature and the future of military logistics itself.

It is patent from this analysis that one of Van Creveld's central arguments – that improvement in military logistics is impossible – is incorrect. That the British and US militaries have improved the efficiency and responsiveness of their logistics since 2003 is beyond doubt. His other major assertion that improvisation is the fundamental characteristic of successful logistics ignores the major impact that meticulously planned, long-term IT systems, new contracting, outsourcing and core-periphery approaches are having on Western military logistics. Even Van Creveld's assertion that forces still rely on supply from the rear and trucks to bring these to the front has been and will continue to be eroded by the introduction of new technologies, the dispersal of combat forces, and the networked and nodal logistics system that is likely to be required to support them. As much as Van Creveld has given to the study of military logistics, this author would contend that such is the nature of recent logistics transformation that much of *Supplying War* is now out of date and conceptually flawed.

Complementing Farrell's work on combat forces, it is clear that the British and US militaries have closely emulated business best practices in transforming their logistics. Following Grissom and Foley et al., it is also clear these top-down emulations of civilian business logistics have been accompanied by bottom-up tactical adaptations. FOBs and CLPs demonstrate how British forces' logistics have also transformed to meet the conditions of the modern battlefield. More broadly, Western militaries' adoption of post-Fordist principles has interesting implications for the future. For one, new technologies such as artificial intelligence, swarm delivery drones, 3-D printing, high velocity distribution from mobility balloons, automated convoys and robotic delivery systems, to name but a few, are likely to further 'challenge the paradigm of the truck'.[104] The RML will continue to evolve with technology and strategic circumstance. Important evidence of this is provided by the recent re-evaluation of military logistics principles in Britain, the US and NATO. For its part, the British Army is currently juggling how to logistically plan for future expeditionary contingency operations and the possibility of major interstate conflict, while simultaneously reducing costs and building robust networks with their industrial bases to provide a potentially strategic edge.[105] Emphasising this point, Carter recently stated that 'the reality is that we can deliver military capability differently if we do so in partnership with industry.'[106]

At the operational level, the future operating environment that the British Army is preparing for will likely involve 'contingency at distance' – the initial rapid deployment of a division or brigade-sized force into an uncertain environment at the end of a potentially stretched supply chain/network.[107] While the renewed focus on the divisional level raises questions as to whether an echelon system would be adopted again during a major inter-state war, the fact that states like Russia increasingly operate 'in the grey' area between conflict and peace would suggest that there is unlikely to be a full return to the days of safe rear zones behind clearly defined front lines. Hybrid warfare will pose a major risk to rearward supply lines, requiring road resupplies to be equipped to fight while also driving the introduction of new delivery technologies. The increasing dispersal of ground forces in response to new technologies will likely further contribute to this. Indeed, the British Army's recent Joint Force 2025 plan to create two 'Strike Brigades' capable of deploying 'rapidly over long distances' and of 'sustain[ing] themselves in the field' (itself a nod to the logistically successful rapid French intervention in Mali), also implies a light logistical footprint enabled by new technologies, contractors, and a greater reliance on local sourcing, rather than a return to the echelon system.[108] Thus, while the wider logistics principles underpinning the echelon system may remain, they will likely continue to inform distributed logistics. Distributed, nodal, logistics seem here to stay. Thus, CLPs and FOBs are unlikely to be historical anomalies, especially as a more resilient defence support network materialises. Conversely, adaptability and flexibility are likely to become even more important as new threats and technologies emerge. Despite the importance of new logistics concepts and systems, it should be recognised that some military logistics principles, such as foresight; agility;

---

104   Interview, senior British Army logistics officers responsible for future doctrine, Andover, 9 June 2015.
105   General Nick Carter, (17 February 2015) comments at Chatham House on 'Future of British Army'.
106   Ibid.
107   Interview, logistics officers.
108   HM Government, *National Security Strategy and Strategic Defence and Security Review*, 10, 6.

co-operation; efficiency; and simplicity are unlikely to change.[109] Moreover, notwithstanding the introduction of MJDI and the adaptation of FOBs and CLPs, as discussed in Chapter 6, at the tactical level many logistics practices have been relatively untouched by post-Fordism, which as this chapter has shown, has predominantly transformed logistics management and structures. Van Creveld remains right in some respects.

Crucially, however, it is important to recognise that the wider force structure in which FR20 has been designed, through a complex, outsourced, rotational system, is a profoundly post-Fordist solution to these old problems. This re-structuring of Britain's land forces around post-Fordist principles also raises interesting strategic implications. By adopting the 'Total Cost' approach to solve the demands of less supply of, but potentially equal or greater demand for, ground forces, the plans' leveraged nature means that it lacks the structural flexibility to respond to strategic shocks. Army2020's rotational system is simply not designed to deal with these kind of shocks; there is very little slack in the system at present, certainly not enough to allow the army to conduct two simultaneous medium-sized operations, as it did in Iraq and Afghanistan between 2006–2009, nor crucially, to conduct division-plus operations.[110] It is arguable therefore that Army2020 is based on the central strategic assumption that there will be enough warning time before a major conflict that requires an army of hundreds of thousands, rather than tens of thousands, of troops. To generate a force this size would require conscription, and these conscripts would likely be trained by cadres of regulars not engaged in fighting.[111] In this context, similar to Haldane's plans, the Army Reserve would likely reinforce the rest of the regulars to buy time for conscripts to be trained. In the opinion of one of FR20's earliest proponents, Lieutenant General Sir Graeme Lamb, it is for this reason that the reserve component remains potentially very important: 'What the reserve gives you which the regular cannot do is scale … so if you've got the need to enlarge on the unexpected, truth of the matter is, scale sits with a reserve, not the regulars.'[112] As such, at the strategic level, Army2020 and FR20 are, by their very design, limited. Even with the integrated reserves component, and the availability of an ex-regular-reserve of 35,000 troops with at least three years' experience, Army2020 essentially provides a core around which a much larger conventional army required for national defence could consolidate. Decisively, it offers the potential for cheap, scalable mass. Interestingly, this indicates an acceptance of the benefits of a more Fordist mode of generating military capability in the event of a national emergency. Indeed, it suggests that the British army is cognisant of the limits of post-Fordist-based force structures in a major conventional near-peer war.

109   British Army, *Tactical Logistics Handbook*, 11.
110   Comments by Lord Dannatt, Global Strategy Forum.
111   Interview, Lieutenant General Sir Graeme Lamb, 13 July 2015. Henceforth Lamb.
112   Interview, Lamb.

In *The Sources of Military Change*, Chris Demchak considered the conceptual, systematic, and organisational implications associated with the integration of the information technologies associated with the RMA into Western forces' doctrine. For Demchak, this IT-enabled systematic transformation had the potential to make combat forces more fragile rather than robust. Demchak argued that 'the long-term structural effects of the emergent worldwide change in military organisation, based on information technology, are not well understood', and that operational effectiveness depends on fewer surprises. This in turn is reliant on less complexity, greater advanced knowledge, and better responsiveness through redundancy. But modern RMA systems enable a way of operating in opposition to these requirements. Thus, Demchak argues there is a 'poor systems fit' between modern military organisation and reality, and she places the blame for this on the failure of the military to adopt a network approach needed to underpin advanced IT use, and on the 'strongly US tendency to inappropriately and incompletely transfer private firm lessons to public task environments.' As a result, Demchak argues that militaries that have transformed around RMA principles will be prone to the 'emergent surprises' associated with complex systems. Demchak's analysis is highly interesting as it not only represented an early attempt to understand the systematic impact of the RMA, it also tied the adoption of its principles to the military's emulation of civilian businesses. However, Demchak only considered combat forces in her analysis, thereby raising questions as to the strategic and systematic implications of the recent military logistics transformation?

Complementing Demchak, at the strategic level, the adoption of JIT processes and the accompanying SCM approach in military logistics has potentially profound consequences for the West. Kane has presciently noted that while JIT 'may be a useful slogan for business management ... it is a dangerous philosophy for defence.'[113] While JIT procedures are cost-effective and efficient, operational effectiveness is the final and deadly standard against which military logistics systems must be judged. Coupled with questions over the impact of JIT on logistics performance there are growing concerns about the nature of the SCM approach to logistics which underpin JIT principles. Recently, Martin Christopher and Matthias Holweg have argued that since 2008, ongoing price turbulence across a number of key market indicators has undermined the basic assumptions of the SCM approach. Crucially, they argue that due to this greater volatility, 'supply chain practices may no longer fit the contexts most businesses operate in – primarily because these practices were developed under assumptions of stability that no longer exist.' Although SCM possesses some flexibility, it does not possess the structural flexibility needed to respond to the major changes in the market, which is occurring in the current era.[114]

---

113   Kane, *Military Logistics*, 155.
114   Christopher and Holweg, 'Supply Chain 2:0', 63.

Whatever the exact relationship between the markets and the strategic situation, it seems clear that since the end of the Cold War the world has become politically, environmentally and economically less stable. Yet, at precisely the time when complexity and uncertainty are increasing, most Western militaries are downsizing while adopting total force concepts and structures in an attempt to maintain capabilities. Clearly, there are advantages to organising integrated forces at tiered levels of readiness, but the re-structuring of these forces is, like the SCM approach the total force concept mirrors, based on relatively stable strategic assumptions of supply and demand. There is little slack left in this more efficient system. Conventional and hybrid threats are increasing and the total force model may not be up to the challenge of meeting them. Indeed, there is already some evidence that Western logistics planners are refocusing on the JIC system in case of another large-scale conventional war.[115] In transforming not only their military logistics systems, but also their entire force structure and readiness around post-Fordist principles, Western militaries are now more vulnerable to strategic shocks that could negate the assumptions on which much of the recent logistics transformation was based. Indeed, the British military's embrace of the Defence Support Network indicates an appreciation of the potential vulnerability of SCM identified by Demchak. The question now is whether a fully networked logistics system, with enough slack and stockpiles to ensure redundancy, can be implemented before the assumptions underpinning SCM are fully tested. However, in the next chapter I turn to examine why a post-Fordist approach came to be utilised in implementing FR20.

---

115   Interview, British Army logistics officers.

Chapter 4

'A Finger in the Wind Thing':
Intra-Party Politics and
Origins of FR20

In Chapter 2, I discussed how economic, strategic and politico-ideological factors have cyclically provided the impetus for past periods of reserve reform, and how these attempts to transform the reserves were heavily curtailed by organisational friction and resistance. I also argued that the sources of reserve reform have usually been primarily political rather than military. In the last chapter, I examined how the post-Fordist approach has changed how military logistics is delivered and influenced wider force structure designs. Here, I want to build on those arguments by examining how the current FR20 transformation originated and how it was implemented. Following Allison, in this chapter I focus on how and why the most recent debate over the position of the TA in British defence was influenced by the desire for economies in defence, strategic uncertainty, and most importantly, by various political and military stakeholders. Contrasting Edmunds et al.'s opinion that 'the most important long-term driver for change [in the reserves was] strategic in nature', I argue that the intensely intra-party political origins of FR20, and the army's resistance to the policy, are of critical importance to understanding the evolution and implementation of FR20.[1] Building on this analysis, the chapter then charts how these origins, coupled with other organisational frictions and personal tensions, have caused FR20 to be tested and adjusted at each step of its development and implementation, resulting in important revisions to the policy. Ultimately, I argue that these intra-party political origins and the army's resistance to them, meant the policy developed ad hoc, thereby lacking coherence and causing organisational issues which had not been foreseen.

The Context of FR20

As Strachan has noted, the purpose of history is not simply to tell us what is similar to the past, but also what is different. With this in mind, three major

---

1   Edmunds et al., 'Reserve forces', 120.

contextual differences between the previous periods of reserve reform and that of the current period should be stressed: the impact of the global recession on British defence spending; the strategic uncertainty of the 21st century; and the post-Fordist principles discussed in the last chapter that Western militaries have utilised to adapt to these pressures. In terms of the impact of economics on FR20, Paul Cornish and Andrew Dorman have observed how the stark economic reality of the 2010 Spending Review heavily influenced the National Security Strategy (NSS) and the Strategic Defence and Security Review (SDSR) of that year. Taken together, these started the transformation of Britain's military just as it started to drawdown from over a decade of involvement in Iraq and Afghanistan. Significantly, Cornish and Dorman detail how the economic context of the reviews dictated that fiscal security was the central premise upon which the SDSR rested. As they claim, this contention remains open to debate given nations' ability to access international money markets; in reality the SDSR was 'politics-led' and representative of the Conservatives' ideological views on the economy.[2] Senior officers who implemented the ensuing defence cuts have argued that despite the austerity rhetoric surrounding them, they were ultimately 'a political choice'.[3] As both the NSS and SDSR occurred concurrently with the 2010 Spending Review, and were completed in only five months, some have argued that the SDSR was 'a treasury-led defence review', highlighting how policy formulation was hurried and improvised at the time.[4] In 2014, the Defence Secretary acknowledged 'that Army2020 was designed to fit a financial envelope' and that financial considerations 'took primacy over the country's abilities to respond to the threats, risks and uncertainties contained in the National Security Strategy'.[5] The arguments offered below support this further. Nevertheless, this primarily politico-ideological context was also complemented by strategic and organisational factors.

The argument can be made that the strategic situation in the 21st century is qualitatively and quantitatively different to any other period in history due to the impact of globalisation and the spread of communications technology.[6] Numerous sociologists have argued that we are now in a period of 'late modernity' that is fundamentally different to the classical modern period of the

2   Cornish, P. and Dorman, A. (2011) 'Dr Fox and the Philosopher's Stone: the alchemy of national defence in the age of austerity', *International Affairs*, 89(5), 346.

3   Interview, Major General Kevin Abraham, 14 January 2015; interview, Major General Dickie Davis, 27 February 2015, henceforth Davis.

4   Cornish, P. and Dorman, A. (2013) 'Fifty shades of purple? A risk-sharing approach to the 2015 Strategic Defence and Security Review' *International Affairs*, 89(5), 1183.

5   House of Commons Defence Committee [HCDC] (2014) 'Future Army2020, Ninth Report of Session 2013–14, HC 576' London: TSO, 5.

6   Giddens, A. (1990) *The Consequences of Modernity*, Cambridge: Polity; Held, D., McGrew, A., Goldblatt, D. and Perraton, J. (1999) *Global Transformations* Cambridge: Polity Press.

Cardwell and Haldane reforms.[7] Despite critique of the 'global village myth' that underpins much of the strategic implications of this change,[8] it is clear that strategic uncertainty and the flexibility of military forces required to cope with globalisation remain key assumptions of British defence policy.[9] It is also clear that the range of tasks being assigned to the military since the end of the Cold War has increased vastly, with conventional war-fighting duties; peace support operations; counter-insurgency; capacity-building abroad; anti-terrorist and aid to the civil power at home all key tasks for the British Army. While there are of course always strategic uncertainties – the debate over the 'blue water' strategy in Haldane's time is just one example of this – for the first time it is the multitude of possible threats and tasks that provided another supporting rationale for the current transformation. Similarly, the impact of the wars in Iraq and Afghanistan, especially in terms of public adversity to long-term interventions and distrust of political 'spin' remain important historical differentiators. While there is some continuity with the strategic contexts of the past, FR20 has therefore been influenced by different strategic problems to its forebears, as well as different intra-party and intra-service dynamics.

Background to FR20

The three major periods of reserve transformation examined in Chapter 2 has already revealed that Britain's reserve army is resistant to reform and that they failed to provide a workable system of overseas deployment excluding existential wars. Thus, the TA and its antecedents have been a strategic rather than an operational reserve. FR20's origins start with the Options for Change defence review of 1990 that cut the TA's establishment from 76,000 to 63,500.[10] This decrease created structural problems within the TA, whilst also leaving it without a clear role. According to one British general, during this period the absence of a defined role led to a 'system of individual backfills [being] introduced which would last for the next 20 years'.[11] Lacking its Cold War reinforcement role, reservists were assigned piecemeal to regular units and usually deployed as individual augmentees. Meanwhile, the greater operational tempo that the military experienced under Prime Minister Blair revealed inadequacies in

---

7   Giddens, *The Consequences of Modernity*; Bauman, Z. (2006) *Liquid Times: Living in an Age of Uncertainty*, Cambridge: Polity.
8   Porter, P. (2015) *The Global Village Myth*, London: Hurst.
9   HM Government (2010) *Securing Britain in an Age of Uncertainty: The Strategic Defence and Security Review*.
10 Taylor, C. (2010) 'A Brief Guide to Previous British Defence Reviews', London: House of Commons Library.
11 Interview, Davis, 27 February 2015.

TA training quality. A lack of funding also kept it short of critical equipment. Coroners' investigations into TA soldier deaths in Iraq emphasised these issues; their conclusions underscored the legal requirement that reservists be as well-trained and equipped as possible.[12] In terms of the TA's future deployment, this meant that the 'legal implications [of deploying insufficiently trained and equipped reservists] were huge'.[13] These structural and legal problems provided some of the impetus for FR20.

The army began to address these TA shortcomings in 2004. Yet, wider structural and functional problems remained a low priority with both organisations committed to operations in Iraq and Afghanistan, where 25,000 reservists served.[14] By 2005, the TA was at its weakest strength since its foundation, reiterating the paradox that deploying the TA to conflicts other than national emergencies often resulted in signoffs and lower recruitment that threatened the organisation's future.[15] As a result, the Chief of the General Staff (CGS), Richard Dannatt, ordered the army to undertake a comprehensive review in late 2008. This proposed three potential courses of action to decrease the TA's strength to between 24,000 and 8,000.[16] It is important to stress that this was an internal Army review; it was not discussed as part of a wider debate within British defence ahead of the forthcoming SDSR. Thus, according to another former CGS, when it became clear that the MoD was considering alternate plans – 'all work stopped on the army's reserve plan in 2009'.[17] Although it was still active in Afghanistan and had a fully trained strength of around 19,000, 'as an organisation [the TA] was in stasis, wondering what was going to happen next'.[18]

By October 2009, there was a new CGS, David Richards, and with the impact of the 2008 financial crisis reverberating through government, the MoD temporarily halted TA training in an attempt to save £20 million.[19] Meanwhile, senior TA officers became increasingly concerned that the Army wanted to radically cut their organisation. They viewed the suspension of TA training as proof that the army's leadership was prepared to allow the organisation degrade in order to rationalise further cuts.[20] This view of a calculatingly malicious army policy to degrade the TA to the point where it could then be reformed – but on the army's terms – gained purchase with the TA's Parliamentary supporters. The importance

---

12 *BBC News*, 30 January 2007, 'New inquest sought for TA soldier'.

13 Interview, General Sir Peter Wall, 14 January 2015, henceforth Wall.

14 Interview, Davis.

15 *The Daily Telegraph*, 13 July 2008, 'Territorial Army soldiers to be ordered to fight in Iraq and Afghanistan'.

16 *The Daily Telegraph*, 28 September 2008, 'Territorial Army faces deep cuts'.

17 Interview, Wall, 14 January 2015.

18 Ibid.

19 *The Daily Mail*, 10 October 2009, 'Territorial Army ordered to halt training as pressure mounts on budgets'.

20 Interview, former regular officer, 9 September 2015.

of ending the neglect of the reserves as a rallying call for transforming the TA is hard to exaggerate. Impetus for FR20 would come more from an increasingly vocal and politicised TA lobby of prominent serving and veteran reservists, members of Parliament and government ministers, than from the Army leadership. From its outset then, the transformation exhibited both externally-imposed and intra-service dimensions.

Policy Exchange Paves the Way?

FR20's seminal moment came in September 2010, when the think-tank Policy Exchange unveiled a report intended to influence the hurried SDSR process. *Upgrading Our Armed Forces* – authored by former Director Special Forces, Lieutenant General Sir Graeme Lamb, and former 22 SAS commander, Richard Williams – argued that the MoD seemed 'to be a reluctant user of its reserve forces'. In response to the threat the Dannatt review posed to the TA, the report argued that it should be expanded by doubling its establishment to 60,000. Indeed, Williams and Lamb drew attention to 'a tendency within the MoD to cut/limit their [reserve force] numbers or starve them of resources as a way of funding investment in the standing forces.' By comparing with other nations' reserve forces, they also condemned using individual TA soldiers to backfill regular units on operations,[21] whilst noting how a reservist cost between one quarter to one fifth of a regular. Williams and Lamb thus called for a 'significant mind-set shift within the senior [regular] leadership of the military', and a 'strategic shift in the way that reservist and regular manpower is managed' to reinvigorate and re-orientate the reserves to meet the demands of the impending SDSR.[22]

Although Policy Exchange delivered a coherent argument around which advocates of reserve transformation could coalesce, the political impetus to review the future role of the reserves was growing before then. On 21 July 2010, the Commons' Defence Select Committee questioned Defence Secretary, Liam Fox, on issues related to the SDSR. During this session, Conservative backbencher Julian Brazier questioned Fox and senior MoD civil servants on whether or not the Reserves' cost effectiveness was being considered in the review.[23] Having served with the reserve 21 SAS, Brazier had deep knowledge of the TA and been its ardent advocate in Parliament for over 25 years. Following the session, the Committee voiced frustration that the MoD had not conducted a study on investing in the reserves as part of the SDSR, especially a cost/benefit

---

21 Williams, R. and Lamb, G. (2010) *Upgrading Our Armed Forces*, London: Policy Exchange, 47.

22 Williams and Lamb, *Upgrading Our Armed Forces*, 48, 51, 53.

23 House of Commons Defence Committee (2010) *First Report, The Strategic Defence and Security Review*, London: TSO, 26–28.

analysis. It therefore recommended 'that the increased use of Reservists should be properly covered by the National Security Council (NSC) in its discussions'.[24] Within the context of austerity, the apparent cheaper cost of reservists vis-à-vis that of regulars provided another important justification for re-examining their role in British defence.

The first of the NSC meetings to agree the detail of the SDSR was to be chaired by Cameron on 28 September. Perhaps with this timeframe in mind, many of Williams' and Lamb's recommendations were published in *The Times* on 15 September. Both also appeared before the Defence Select Committee that day. Clearly, at least some of the Committee were already predisposed to their arguments. A day later Williams stressed the TA's 'pay-as-you-go capability' in another article.[25] The next day, Whitehall sources reported 'very, very strong tensions developing' in the MoD between some ministers and the Army about the TA.[26] By 28 September, both major newspapers were reporting how the Policy Exchange document was causing frictions in government over the TA's future.[27] Richards labelled these a 'classic inter-service battle' in which 'each service defended [their] respective turfs'.[28] When Richards was made CDS in October 2010, his successor, General Sir Peter Wall, also regarded the regular-reserve issue as a 'zero-sum game', but more in terms of intra-service organisational survival whereby either the Army or the TA would be worse off.[29] Thus, in post-Fordist terms, Dannatt, Richards and Wall all sought to preserve the Army's 'core' capabilities and regarded the reserves, as 'not as important as perceived' by others.[30]

Although FR20 gained cross-party support when it was unveiled, it is essential to understand its ideological and politically-charged origins within the Conservative Party. Policy Exchange, founded by Francis Maude MP in 2002, is one of the most influential think-tanks on Britain's political right. It had ties to David Cameron and is funded by donations from some of the biggest donors to the Conservative Party. At the time, Cameron was in the early days of his premiership in the coalition government. Cameron's alliance with the Liberal Democrats had left him vulnerable to censure from the right of his party, which, amongst other grievances such as Britain's EU membership, was disinclined to accept defence cuts. Similarly, his relationship with his Defence Secretary, Fox,

---

24   Ibid., 16.

25   *The Times*, 15 September 2010, 'Only hi-tech forces can win wars of the future'.

26   *The Times*, 16 September 2010, 'Fighting force of the future needs twice as many part-timers, say ex SAS chiefs'.

27   *The Daily Telegraph*, 28 September 2008, 'Territorial Army faces deep cuts'; *The Guardian*, 28 September 2010, 'Territorial Army in sights of defence review'.

28   General Sir David Richards (2014) *Taking Command*, London: Headline, 295, 297.

29   Interview, Wall, 3 August 2015.

30   *The Times*, 17 September 2010, 'Battle brews in Whitehall as Tory MPs push to increase the Territorial Army'; Richard Dannatt comments at Global Strategy Forum, 14 July 2015.

was troubled as Fox had run in the 2005 Conservative leadership contest. Fox had clashed with Cameron on numerous political and defence issues, and as a senior figure on the party's right had considerable backbench support.[31] Pro-reserve Conservative MPs also had his ear at this time. In September 2010, the danger of a coalition of Fox and right-wing backbenchers delivering a political blow to Cameron over the issue of defence cuts was real, especially as he could not give ground on an EU referendum due to Liberal Democrat leader Nick Clegg's veto on the matter.[32]

It is within this intra-party political context that the Policy Exchange recommendations were supported by Brazier and David Davis, another former 21 SAS reservist. Both were parliamentary champions of the TA and backbenchers on the right of the Tory party, and both MPs distrusted the army's intentions due to its track record of prolonged underinvestment in, and recent plans to reduce, the TA.[33] They could draw on the support of other prominent Tory backbenchers, including Bob Stewart, Julian Lewis, John Baron, and up to 15 others. Inside the TA, this lobby had the backing of one of Britain's wealthiest men and Assistant Chief of the Defence Staff for Reserve Forces, the Duke of Westminster, as well as senior TA officers such as Brigadiers Sam Evans, John Crackett and Ranald Munro, who were concerned about the TA's survival if its fate was left to the army.[34] Crucially, according to one former CGS, these were 'quite independent people, not short of going off on their own political tack ... they want[ed] to be a part of the army and part of a separate political axis, and it was a very powerful political axis'.[35] Indeed, it is noteworthy here that the Duke of Westminster was later appointed as the first Deputy Commander Land Forces, while both Brigadiers Munro and Crackett would eventually be promoted to major general and the two most senior reservist positions. These officers represent Posen's mavericks, with strong links to the organisation's political masters. In the opinion of regular senior officers, these links and lobbying perilously blurred the Huntingtonian civil-military divide. Yet, to senior reserve officers and politicians, this was precisely how one created support for policies. In fact, senior TA officers and politicians would argue that the army's leadership was also playing politics by trying to reduce the TA's size to insignificance, and without consultation. Indeed, the lingering existence of diverse perceptions of each faction's political motives highlights how suspicious they were of each other; the level of intra-service rivalry is palpable.

Although it is not clear if Brazier and Davis – or Policy Exchange – specifically asked Williams to examine the reserves, the report was certainly not drafted in

---

31 *The Financial Times*, 29 September 2010, 'On manoeuvres: Fox vs Cameron'.
32 *The Financial Times*, 24 June 2016, 'Brexit: Cameron and Osborne are to blame for this sorry pass'.
33 Interview, Davis, 27 February 2015.
34 Interview 18.
35 Interview, Wall, 14 January 2015.

a political vacuum. In the weeks before the publication of the SDSR, Cameron personally intervened to stop the army from cutting the TA. Media reports at the time stressed Cameron's commitment to his 'Big Society' policy as a central reason for this, and in a subsequent keynote speech to businesses Cameron explicitly linked Big Society with the TA.[36] Supporting his Big Society agenda, the TA would help the army improve its community engagement and presence within British society and, therefore, help reduce the widening 'civil-military gap' between the Army and society since the end of conscription in 1960.[37] With its emphasis on increased volunteerism the Big Society political narrative supported the growth of the reserves.

Meanwhile, news reports warned against reducing the TA, with a Whitehall source commenting: 'The TA is unfinished business – they should have been restructured and cut before now, but a lot of them are well-connected and eloquent and they're very good at lobbying.'[38] When the SDSR was published on 19 October 2010, it was obvious that the Policy Exchange report, Tory backbench agitation, the TA's lobbying efforts, the Defence Committee's criticisms, sympathetic media coverage, and finally Cameron's intervention, had shaped the reserves' destiny, as it stated that a review of the reserves would be undertaken despite the army facing major budget reductions and reorganisation.[39] This included a restructuring of the army around a new 'Future Force 2020' (FF20) model, with a brigade-based, higher and lower readiness, rotational force structure aimed at conducting limited contingency operations. FF20 thus clearly followed post-Fordist principles, with core and periphery forces organised around networked brigades, and a renewed desire to examine outsourcing capability to reservists.

The intra-party political stimulus for a reserves review is reinforced by Lamb's assertion that, 'the Policy Exchange pressure from Parliament and back benchers were the reasons why [Cameron] then looked at actions that were taking place within … the army … and therefore said "we should have a review"'.[40] Similarly, according to Wall, despite the army leadership's clear opposition, the 'reserve thing [was] politically imposed … Cameron had to give ground to some parts of the Tory backbench and the reserves was a way of doing it … Essentially it was a political fait accompli and we just had to get on with it'.[41] FR20's intra-party political and ideological origins, rather than in the

---

36   *The Times*, 10 October 2010,'Prime Minister appeals to military chiefs not to cut Territorial Army'; David Cameron, 2 December 2010, 'Business in the Community' speech.

37   Strachan, H. (2003) 'The Civil-military "gap" in Britain', *Journal of Strategic Studies*, 26(2), 43–63.

38   *The Times*, 11 October 2010, 'Vital Territory'.

39   HM Government (2010), *The Strategic Defence and Security Review*, 15.

40   Personal communication, Lamb, 1 June 2015.

41   Interview, Wall, 14 January 2015.

military/strategic spheres have been confirmed by other senior officers and a former defence minister.[42] These external, ideological and intra-party political drivers are critical to understanding FR20's ensuing development. They not only created tensions between Conservative backbenchers and ministers, and between government politicians and the army, but also between senior officers in the TA and the army, impacting FR20 at almost every step in its evolution. Perhaps most significantly, like the Cardwell-Childers and Haldane reforms before it, FR20's political origins would also create dissonance between its vision for, and the reality of, its organisational outcomes.

An Independent Reserves Commission?

Following the publication of the SDSR, Cameron briefed Parliament that an independent commission would be created to study the reserves. Lamb has recounted both the hurried and ad hoc development of this policy decision:

> So ... then I got a call, literally in my garden, from ... [Edward Llewellyn] the Prime Minister's Chief of Staff, who said: "You're the only name that's come out of the hat who we'd trust to play this straight. But we're going to put a commission up, General [Nick] Houghton is going to be the serving [member], Julian Brazier is going to be the MP, and I'd like you to be in, and the Prime Minister is walking across to Parliament [to announce it], will you do it?"[43]

The quote is interesting as it demonstrates the degree to which the reserves issue had become politically disputed even at this early stage. Not only was the Prime Minister's Chief of Staff personally involved in enlisting an independent commission member, he was also reiterating the need to 'play it straight'. This provides further evidence of the deep tensions between politicians and senior TA officers on the one hand and senior army officers on the other.

The inclusion in the Commission of Brazier, the reserves' most vocal Parliamentary supporter; Lamb, whose reserve views opposed those leading the Army; and Houghton, who was in the running for the position of CDS, does raise questions about the Commission's impartiality. Wall's opinion is that the review was exposed to 'significant pressure', with the TA lobby 'banging this drum ... in a sense coercing Houghton's Commission to agree bigger numbers ... [and] if we're really honest, slightly to political appetite, slightly to political order, [Houghton] said: "yeah they'll get to [the trained strength target of] 30,000 fine, it's a tiny

---

42 Interview, Davis, 27 February 2015.
43 Personal communication, Lamb, 1 June 2015.

proportion of the national workforce"'.[44] As I will show in the concluding chapter, this forecast would prove to be deeply flawed.

In July 2011, the Commission reported back to Parliament, confirming that the UK's reserve forces were in 'severe decline' and formed too small a part of the nation's military capability. It recommended a number of major changes to their role and structure, with a particular emphasis on the TA, which it recommended be increased from 19,230 to 30,000 by 2015, stressing the need to 'commit to returning formed sub-units to "the fight"'.[45] Other important suggestions included the pairing of regular and reserve units and the integration of the TA into the FF20 force structure, and a number of employment and welfare reforms intended to improve recruitment and retention. The political spotlight had found the TA, and the stage had mainly been set by the intra-party dynamics within the coalition government, especially the alliance between TA-supporting Tory backbenchers on the right of the party and serving senior TA officers, and their championing of the Policy Exchange reserve recommendations. These combined to put considerable pressure on the army to implement a reserves transformation programme that its leadership deemed peripheral. This evidence supports Zisk's contention of the importance of constructing alliances within the policy arena in order to execute externally-imposed innovation that benefits a particular group's organisational interests.

Paired Fates: The Regular Army and Future Reserves 2020

Faced with a further budget reduction in May 2011, the MoD began conducting a three-month review on how its new fiscal targets would impact FF20. In July it concluded that the SDSR 'was not an affordable proposition' and that the Army now needed to be shrunk from 102,000 to 82,000.[46] Meanwhile, on 3 July the Independent Commission released its report. Events now moved quickly. With reports of much political tension between Cameron and Fox over the exact character of the cuts and amidst 'considerable disquiet' between senior Regular officers, on 15 July the Cabinet took the cost-saving decision to reduce the army and invest in the reserves.[47] This decision was heavily influenced by the almost simultaneous conclusion of the 'three-month review exercise' and the recommendations of the Independent Commission. Given the political tensions over the defence cuts, combining both reports presented Fox – and by extension Cameron – with a propitious opportunity for decreasing the army while simultaneously increasing the reserves.

---

44   Interview, Wall, 14 January 2015.
45   House of Commons Debate, 23 April 2013 c286WH; *The Independent Commission*, 30, 26.
46   Interviews, Wall, 14 January 2015; Davis, 27 February 2015.
47   Ibid.; *The Daily Telegraph*, 19 July 2011, 'MoD sacrifices manpower to pay for equipment'.

Taking the recommendations of both reports was politically beneficial in that it allowed Cameron and Fox to quell Conservative backbencher and Labour criticisms of their defence policy by presenting the redevelopment of the TA as fair recompense for reducing the army. Moreover, both plans could be couched in the austerity-inspired language of increasing efficiencies and reducing costs. On 18 July, Fox briefed Parliament that the Army needed to be downsized and restructured. However, he stressed that any reductions would be offset by a £1.5 billion investment in the reserves. This would increase the trained strength of the TA from 19,230 to around 36,000 by 2020. Fox stated that, 'if the [TA] develops in the way we intend, we envisage a total force of around 120,000, with a regular to reserve ratio of around 70:30'.[48] According to a Defence Select Committee report, such numbers were presented to the CGS by the Permanent Secretary at the MoD without prior consultation.[49] Furthermore, according to Wall, by including 8,000 untrained reservists in his figures, Fox was able to claim that:

> the Army was the same size, just the composition was changing ... which was an obfuscation and a deliberate lie ... it wasn't a surprise to any of us [in the Regular Army] that, slightly fallaciously, the government had sought to portray the increase in the Reserves as a fair compensation for the reduction in the Regulars.[50]

According to another general, Fox's decision to blend both reports' recommendations to create a new integrated regular-reserve force structure, was 'ad hoc' and politically opportunistic.[51] Certainly, as the leader of the organisation subject to such deep cuts, Wall believed that 'The reserve thing was politically-imposed in terms of it being a political motive'.[52] This perception that FR20 was ideologically motivated and politically opportunistic would influence its subsequent development and implementation. From the outset, there were tensions between the presentation and reality of FR20.

Now that FF20 was to be implemented with a much reduced regular component, the army 'quite quickly realised [it] needed to set up a design team that was outside the chain of command ... that this wasn't a "business as usual" proposition'.[53] In May 2011, then Lieutenant General Nick Carter, assisted by then Brigadier Kevin Abraham, set out to design a new model for the army that would become known as 'Army2020'. Army2020 was tested at each stage in its development and was consciously designed to integrate

---

48 *Hansard*, 18 July 2011, col. 644.
49 Defence Select Committee, *Future Army 2020*, 5, para 32.
50 Interview, Wall, 14 January 2015.
51 Interview, Davis, 27 February 2015.
52 Interview, Wall, 14 January 2015.
53 Ibid.

the reserves. This inclusion of the TA represented an awareness that reserves transformation would occur – that the politicians had triumphed. Echoing post-Fordist principles, Army2020 divided the organisation into a high readiness 'Reactive Force' and a lower readiness 'Adaptive Force' for follow-on operations. Within both forces was a 36-month operational readiness cycle, whereby brigades in the Reactive Force, and units within the Adaptive Force would come up to readiness for potential deployment for 12 out of every 36 months. If deployed, another system known as the 'harmony guideline' designated that forces at readiness could expect to be deployed once in each five-bloc cycle; in reality for about six of every 30 months. Critically, the less-ready Adaptive Force was to provide the greater capability in roulements four and five of a deployment. Unveiling the plan, Carter described it as: 'new and imaginative and original ... Getting there will be challenging ... And none of this happens very quickly, it will be a gradual process.'[54] Most decisively, supporting the 'Whole Force Concept', Army2020, detailed a more important operational role for the TA in the Adaptive Force, which aimed 'to deliver a genuinely useable and capable reserve that is integrated with paired regular units'.[55] This move significantly increased the operational and training requirements on the reserve component and underscored the post-Fordist principles underpinning Army2020, and especially the logistics component of the plan.

Failure to Adapt: Organisational Reality Bites

Although the army was now clear on its future structure, and how the reserves fit into it, according to a former CGS detailed Reserves planning:

> Was being handled much more by the department [i.e., the MoD] than the army, because it was an externally proposed proposition that had never been fully tested with us ... It was a finger in the wind thing ... There was no science behind it ... there was no evidence it could be done. And there was no thought about if you decided to do it how you would actually go about it.[56]

---

54 House of Commons (2012) *Library Report* 'Army2020', 6.
55 MoD, *Transforming – An Update*, 4.
56 Interview, Wall, 14 January 2015.

Richards supports this claim, detailing how 'the motor for this project was Houghton's team, operating largely outside the [army] process'.[57] The fact that the planning for the reserves occurred without the rigorous testing that Army2020 underwent, and partially outside the army, is significant as it highlights the level of suspicion between senior army officers and their political masters on the issue. Further problems lay ahead.

In November 2012 the MoD published its Green Paper, *Future Reserves 2020: Delivering the Nation's Security Together.* This supported the increase in TA trained strength to the 30,000 recommended by the Commission, coupled with a much more significant role for the reserves and their full integration into the Whole Force. However, the date set for this by Hammond at the time was 2018, not 2015 as the Commission had recommended, nor the 2020 deadline that the army appear to have understood.[58] Again, this lack of clarity indicates the friction and confusion in the evolution of FR20. The paper pledged to invest £1.3 billion over the next ten years, increasing the amount recommended by the Commission by £300 million. It also contained a comprehensive list of reforms that would be undertaken in order for the reserves to meet the requirements laid out in the SDSR and Army2020. These included the propositions to rename the TA the Army Reserve to reflect its more integrated role; investments in training and equipment; extended mobilisation powers; increased reservist remuneration and welfare packages; and better engagement with reservist employers.[59]

Crucially, the Green Paper stressed that there would be a 'change from using the reservist on an individual basis to mobilising formed sub-units.'[60] It proposed a 15 per cent increase in the annual training requirement to ensure units could deliver collective tasks 'at the platoon, company and battalion' levels. Indeed, the deployment of formed sub-units was mentioned 14 times, and became a central tenet of FR20.[61] Hammond also stressed the importance of deploying sub-units, announcing that 'this transformation of the reserves will see a radical shift in the way in which we use them'.[62] This emphasis on formed reserve units was complemented by a similar focus on their routine use, which was stressed 16 times, highlighting how the vision for the Army Reserve had changed from a force of last resort to one that was integral to the army's deployment plans. Nevertheless, notwithstanding these aims, much of the Green Paper's emphasis fell on proposed

---

57 Richards, *Taking Command*, 299.
58 *Hansard*, Commons Daily Debate, 8 November 2012, col. 1026. Wall states 2020 in *The Times*, 24 February 2012, 'The day of the "citizen soldier" has arrived'.
59 *Future Reserves 2020.*
60 Ibid., 6, 16, 27.
61 Ibid., 16.
62 'Consultation launched on the future of Britain's Reserve Forces' available at https://www.gov.uk/government/news/consultation-launched-on-the-future-of-britains-reserve-forces, retrieved 28 July 2016.

changes to mobilisation legislation and terms of service, instigating employer/ family support initiatives, and in increasing monetary compensation to reservists to boost recruitment and retention. As Edmunds et al. have noted, what the Green Paper was recommending was essentially transactional in nature; a change in the readiness and utility of the reserves in order to generate efficiencies, in return for increased investment and support.[63]

As FR20 planning continued, the organisational difficulty of deploying reserve sub-units with the regulars became clearer. The training differential between regulars and reservists meant that the legal responsibility to ensure reservists were 'accredited, regulated and subject to legislation' underpinned any ability to deploy them, supporting Edmunds' argument about the increasing prominence of risk management in British civil–military relations.[64] While institutional rivalry and cultural suspicion also played a role, the army correctly reasoned that deploying reserve units to conflict environments alongside regular units without providing the similar, time-intensive training of the latter had major legal implications. Integrated collective training, therefore, became the crucial first step toward building sub-unit operational capability. Meanwhile, it became clear that outsourcing logistics capability to reserve units required a huge re-organisation of numerous units, with fully qualified reservists in some units being forced to re-train in the newly-required capability. To do so would take much longer than the FR20 schedule allowed. Thus, the organisational realities of reserve service constrained FR20 from the beginning, emphasising how the organisation struggled to adapt to meet the demands of the innovation.

Indeed, the drive for deployable sub-units had been revised and diluted due to these organisational realities when the FR20 White Paper was published in July 2013. Hammond stated that the army, '*while continuing to deploy individuals*, will have a greater reliance on [reserve] formed sub-units and units' (authors' emphasis).[65] The explicit reintroduction of individual backfilling was very important as it lessened expectations that the reserves would deploy as units, which had been a central and continuous strand of reserve transformation discussions since 2010. Crucially, it confirmed that the Army Reserve (AR, the newly renamed TA) would continue to contribute to operations as it had in the past. Routine reserve deployment was another central tenet of FR20. Introducing it, Hammond stressed that, 'Under our new model, the use of the reserves is no longer exceptional or limited to times of imminent national danger or disaster, but is integral to delivering military effect in almost all situations' while also emphasising the 'greater efficiencies in training and equipment resulting from formal pairing between regular and reserve units'.[66] While the Commission had recommended pairing regular and reserve units, the

63   Edmunds et al., 'Reserve Forces'.
64   *Future Reserves 2020*, 22; Edmunds 'British civil–military relations'.
65   *Hansard*, 3 July 2013, vol. 565, col. 924.
66   *Future Reserves 2020*, 7; *Hansard*, 3 July 2013, vol. 565, col. 924.

Green Paper only mentioned it twice. Yet, the call for pairing appeared in the White Paper 16 times. There had clearly been another change in emphasis, caused by the realisation of how the few previous reserve sub-unit deployments were managed successfully.[67] This evolution of FR20 is instructive, because it highlights how the AR's intrinsic organisational realities caused the policies' main objectives to be revised. Decisively, the White Paper recognised that the routine deployment of formed sub-units relied on collective training and close relationships with regular units, which would take time to deliver. This revision emphasised a recognition within government that implementing FR20 would not be as straightforward as originally envisaged.

The Politics of Numbers

With a new role defined, the need to expand the AR came centre-stage. Senior officers questioned the AR's ability to recruit to establishment, while Brazier and other politicians reported 'horrifying' problems with the recruitment process, which had been outsourced to management firm Capita in line with post-Fordist principles.[68] Meanwhile, the scope of Army cuts, and the new reliance on the AR to deliver key capabilities, drew substantial media coverage and, in the eyes of senior army officers, obsessive Parliamentary scrutiny.[69] By putting the AR at the centre of British defence policy, FR20 ensured that its implementation would remain a politically-sensitive issue for Cameron, with backbenchers, the Labour opposition and even Fox (after he was forced to resign) subsequently warning about reducing army numbers before the AR had reached its establishment.[70] Indeed, by linking Army2020 and FR20, the success of the entire Army transformation now hinged on AR recruitment. According to Fox, politically it had become 'a numbers game … and we'd taken an enormous gamble with [those] numbers'.[71] Indeed, Richards also states that Houghton had told the NSC that FR20 needed testing to prove its feasibility, but that 'the government decided to push it through without this sensible precaution'.[72] This echoes Wall's view that the reserves' enlargement was 'not grounded in military experience, military fact, or any credible evidence', and Dannatt's opinion that it 'was based on hope rather than any science'.[73] Indeed, Richards records that Houghton felt outmanoeuvred by the politicians and that his

---

67 Connelly, 'Cultural Differences'.
68 *The Times*, 19 February 2013, 'New Model Army'; *The Times*, 5 November 2012, 'Red tape chokes Army's vital recruitment drive'.
69 *The Times*, 6 July 2012, 'Army cuts take 'military gamble' by placing burden on Reserves'; *The Daily Telegraph*, 8 July 2012, 'TA can't recruit enough "quality troops" for plans'.
70 *Hansard*, 8 November 2012, cols 1025–1034.
71 Interview, Liam Fox MP, 28 May 2015.
72 Richards, *Taking Command*, 299.
73 Interview, Wall, 14 January 2015.

'good nature has been taken advantage of'.[74] Thus, the intensely political nature of FR20's origins and the subsequent intra-service rivalry ensured it continued to be an externally driven process that the army did not fully support.

If Cameron hoped that the White Paper would settle the reserves issue he was to be disappointed. Less than a month after its release – and despite a new £3 million recruitment campaign – leaked Army reports showed that the Army Reserve's strength had dropped by five per cent and that recruitment was 50 per cent below target.[75] Meanwhile, the outsourcing of the AR's localised recruitment system came under continued attack, with Brazier leading the criticism. Rumours circulated that the army wanted FR20 to fail. Despite the fact that media contact with MoD personnel requires prior approval, there were significant leaks.[76] A regular officer involved with Army2020 planning agreed that elements in the army did want the reserves plan to collapse, believing that if it did the political appetite for reducing the regulars would evaporate.[77] The fact that army reports detailing the problems with reserve recruitment were leaked to the media supports this argument.[78] FR20 was becoming an intra-service zero-sum struggle for survival.

By now, the recruitment problem was acute. In March 2012 Capita had signed a £440 million contract to introduce an outsourced, centralised and automated recruitment system. However, in another indication of the ad hoc evolution of FR20, the contract had been negotiated before the policy was confirmed and did not foresee a rapid expansion of the reserves (in another political twist, Hammond blamed Francis Maude – founder of Policy Exchange and now Cabinet Minister responsible for streamlining the civil service – for this oversight).[79] Information technology systems were inadequate, while the paradox that a more deployable reserve required higher medical standards resulted in decreased inflow. The situation was compounded by the fact that in 2010, the new government had banned recruitment advertising. As 2013 continued, the media saw that a keystone of the government's defence policy was faltering and focused intensely on AR recruitment, frequently reporting quarterly manning statistics. The issue came to a head in mid-November when former army officer John Baron tabled a Parliamentary bill calling for a delay to the army cuts until the AR met its recruitment targets. In a sign of the deep divisions within the Conservatives over FR20, Baron received the support of 22 Tory rebel MPs and the backing of most

74 Richards, *Taking Command*, 299.
75 *BBC News*, 11 August 2013, 'Army cuts: Reservists slow to enlist, leaked memo suggests'.
76 *The Daily Telegraph*, 18 November 2010, 'David Cameron: "MoD has a problem with leaks"'.
77 Personal communication, former Regular Army officer, 19 August 2015.
78 *The Daily Telegraph*, 16 October 2013, 'Reforms have left the Army in chaos'.
79 Interview, Wall, 14 January 2015; *The Times*, 17 October 2013, 'Flawed plan to boost TA could put Britain at risk, critics warn'; Interview, Davis, 27 February 2015.

of the opposition. The constituent politics dynamic of the Baron rebellion is also noteworthy: he is a former member of the Royal Regiment of Fusiliers, whose second battalion was to be disbanded in the next round of redundancies. With the impending vote attracting substantial media attention, Hammond stressed that any delay in implementing FR20 would send a 'negative signal' to reservists, and ironically, 'make the whole agenda into a political football'.[80] Yet, the night before the vote Hammond met with Brazier – who appeared at this stage to still favour Baron's position – in a bid to resolve the issue. In this meeting Brazier demanded an annual external audit to monitor FR20 in return for his support for Hammond, which the latter acceded to.[81] It is unclear if Brazier was promised the new Minister of Reserves post at this meeting, but it is noteworthy that within months of coming out in support of FR20 he had been selected for the role. While Brazier's passion made him a natural choice for such a role, one senior officer described Brazier's promotion as a 'political move' designed to give the outspoken backbencher ownership of the reserves problem and, hence, 'shut him up'.[82] Whatever the truth, once Brazier became responsible for FR20 he became much more supportive of it.

The Tory rebellion underscored FR20's mounting political costs. During the following year, intense political and media scrutiny on reserves recruitment continued as numbers failed to rise. In January 2014 the Defence Select Committee raised concerns over the AR's ability to reach its April 2019 manning target; the National Audit Office (NAO) followed suit in June.[83] That same month, the Major Projects Authority (MPA) stated that FR20 appeared unachievable. In July, the first MoD External Scrutiny Team (EST) criticised FR20's implementation.[84] Two months later the Public Accounts Committee echoed the NAO, concluding:

> It is astonishing that the [MoD] went ahead with plans to cut back the regular Army by 20,000 and increase the number of Reservists without testing whether this was doable and without properly consulting the Army itself.[85]

It also noted how the changes to the Capita contract had incurred additional costs of £70 million, considerably reducing the contract's projected savings and hence the rationale of outsourcing recruitment.[86]

---

80 *BBC News*, 20 November 2013, 'Army Reserve rebellion in prospect among Tory MPs'.
81 *The Times*, 20 November 2013, 'Hammond gives ground to rebels before crunch Army vote'.
82 Interview, Davis, 27 February 2015.
83 National Audit Office, Report on Army2020, 6.
84 CRFCA (2014) *Future Reserves 2020: 2014 External Scrutiny Team Report*.
85 Comments made by Nicola Hodge MP, on release of Public Accounts Committee *Report on Army2020*, Eleventh Report, 5 September 2014.
86 Public Accounts Committee, 5 September 2014, *Report on Army2020*.

With both Cameron and Hammond's political capital heavily invested in FR20, Parliamentary questions about the transparency of, and delays in, publishing reserve recruitment data resulted in increased political involvement. Although the 2014 Defence Reform Act removed many restrictions on when and how reservists could be deployed, a further relatively unsuccessful recruitment drive took place in July 2014, and the ensuing major relaxation of the age limit led to ridicule of the AR as a 'Dad's Army' and criticism of FR20's value for money.[87] At this point, Cameron and Hammond sensed that the political risk of FR20 was not diminishing and decided to 'let it go'.[88] By July 2014 Michael Fallon had been appointed Defence Secretary and Brazier reserves minister. One of Fallon's first actions was to adjust the criteria for which military posts counted as AR service, immediately adding 680 personnel to its trained strength. While not unjustified, this statistical sleight of hand provided positive 'evidence' of progress toward the 30,000 trained strength target.

With the political heavyweights divested of FR20, the recruitment problem contributed to the beginning of its revision. The first indication of a potential shift in policy occurred in October 2014, when Carter, recently appointed CGS, appeared to undermine the whole rationale behind FR20, stating:

> A reserve is what it sounds like; it's there for worst-case ... The sense that there is an obligation to be routinely and regularly used is not how I would see this being used. It is there for worst-case. It's certainly not there to mitigate the reduction in regular numbers.[89]

Carter's assertion that the AR would only be employed for national emergencies directly contradicted Fox's, the White Paper's, and Hammond's position that the AR would be used 'routinely' to do tasks that were once the 'exclusive domain of the regulars'.[90] It highlighted the dissonance between FR20's political origins and the organisational realities of the AR, and the army's resultant disinterest in its transformation. Acutely aware of the recruitment challenges, Carter highlighted the reservist's difficulty in maintaining a good balance in the 'equilateral triangle between [the reservist's] employer, his family and himself. What you have to do is explain it's here for worst-case – and keep that triangle absolutely in balance'.[91] This remark offers a clue as to why Carter was now backing away from a more operational role for the reserves. With recruitment figures soon to be revealed as 'shocking', and many of Britain's larger employers concerned that FR20 would

---

87   *The Times*, 14 November 2014, 'Dad's Army: MoD ready to call up the over-50s'.

88   Interview, Wall, 14 January 2015.

89   *The Daily Telegraph*, 28 October 2014, 'Reservists are no replacement for regular troops, head of Army says'.

90   Ibid.

91   Ibid.

radically increase the demands on reservists,[92] Carter's re-appraisal was aimed at both the employer lobby and at enticing more recruits. Carter himself hinted at this, stating there would be a 'refinement' in the army's message to highlight that service would not encroach too heavily on civilian life.[93] Further supporting this, at around this time Carter launched a 'Darwinian' approach to establish the sustainability of unit structures and giving them a year in which to recruit above established strength in order to confirm their sustainability.[94] Despite this, the army also began the cost-saving process of closing smaller AR locations and centralising larger bases following post-Fordist principles, with damaging results on recruitment and retention.

There were other factors at play beyond recruitment. As the new CGS, Carter was unblemished by the previous battles over army cuts and FR20, and as Army2020's chief architect, he had his own political capital. The sustained criticism of the recruitment failures in the media and Parliament gave him a compelling narrative to justify a change of position. Decisively, as Cameron and Hammond had divested their political capital – effectively leaving Brazier to oversee FR20 – there were less political pressures to challenge Carter's decision to change the AR's role envisaged in FR20.

With recruitment and retention figures beginning to rise, by mid-2015 Brazier was confident that 'current strengths are running ahead of schedule',[95] and while progress had been made, politicians and senior officers such as Fallon and Houghton continued to stress that FR20 will take time to implement.[96] Indeed, one senior officer heavily involved in the FR20 process has stated that the timescale for the reforms is unrealistic due to the 'complete underestimation of the neglect of the TA in terms of underinvestment'.[97] Against this backdrop, sustained criticism of the reduction in the Army's size continued after it emerged that it had reduced below its 82,000 target three years faster than anticipated, with media reports of British defence in 'chaos'.[98] The EST stated in its June 2015 report, that: 'Our assessment is that FR20 remains on or near track for delivery' it also remained cautious, while only days later it was reported that the MPA had again given FR20 a red score, indicating that 'successful delivery of the project appears to

92 *The Guardian*, 8 November 2012, 'Firms that discriminate against Territorial Army Reserves may be sued'.

93 *BBC Radio 4* (28 October 2014) 'Reservists are no replacement for regular troops, head of Army says'; World at One Programme.

94 Interview, Carter, 11 May 2016.

95 *Hansard*, Commons Daily Debate, 27 November 2014; 8 June 2015.

96 *The Daily Telegraph*, 1 July 2015, 'Michael Fallon admits government is struggling to recruit Army reservists'.

97 *BBC Radio 4*, 28 October 2014, 'Reservists are no replacement for regular troops, head of Army says'; World at One Programme.

98 *The Sunday Times*, 9 August 2015, 'Redundant troops rehired as army marches into "chaos"'.

be unachievable.'[99] With the reduction in the army complete, but the reserves still badly under-recruited, the sustainability of the Army2020 deployment model, and in particular the Army Reserve's contribution from roule four onwards, remained in question. Indeed, at this time Baron stated the Army2020 and FR20 was 'a plan that has produced a capability gap in the short term and will prove a false economy in the long term and we will live to regret it.'[100] For his part, Fox has also said that FR20 was 'badly synchronised'.[101]

Not that this is surprising given the past attempts to transform the reserves. One senior army officer heavily involved with the reserve recruitment process has stated that the FR20 timescale is 'fundamentally flawed' and, when seen in the context of past periods of reform, is in fact a '20 year transition'.[102] Similarly, Wall has spoken of the need to take a strategic view of the timeline and that the 'army should differentiate between the short-term numbers game, which is a political plan, and establishing [FR20] properly so it stands the test of time and is a system that has the resilience to work well in a crisis.'[103] Unsurprisingly, in the army's view, the pairing of regular and reserve units is central to this, not only in terms of delivering reservist capability but also in terms of fostering better relationships and offering reservists better opportunities. Moreover, the deployable sub-unit requirement still raised questions about politicians' risk appetite for deploying formed reserve units into high threat environments. This is compounded by the fact that reserve commanders are less experienced due to shorter qualifying courses and therefore, conceptually at least, are a greater risk than their regular counterparts. Given these recruitment and risk issues, at an organisational level there are clearly major frictions concerning the implementation of FR20.

Conclusion

Clearly, the ideological rationale of decreased state spending, intra-party politics, and regular-reserve rivalry, not strategy, were the primary drivers of FR20. As this chapter has shown, intra-party politics and ideology played a crucial role in the debates about FR20 and its subsequent organisational evolution. These political origins, located within the backbench of the Conservative Party and closely related to the weak position of Cameron, quickly fused their political goal of reinvigorating the reserves with economic arguments about the need for a cheaper, smaller land force capable of meeting a diverse array of global threats. This in turn led to FR20's

99   CRFCA (2015) *Future Reserves 2020: 2015 External Scrutiny Team Report*; *The Financial Times*, 25 June 2015, 'Army reservist plans "unachievable," watchdog warns'.

100 Ibid.

101 Interview, Fox, 28 May 2015.

102 Interview, Davis, 27 February 2015.

103 Interview, Wall, 14 January 2015.

renewed emphasis on the expansion of the Army Reserves and its integration with the regular army. However, intra-service rivalry and organisational resistance have also profoundly shaped FR20, with strong lobby groups in Parliament, the army, the reserves and the media contesting the rationale and vision for this transformation. At each stage of FR20's evolution there were clear policy changes due to tensions between these groups and the dissonance between desired political outcomes and organisational realities. Indeed, for all of FR20's modernity and professed originality, the fundamental factors curtailing reserve reform – that of recruitment, budgets and strategy – appear to have remained remarkably constant. However, it is clear that in the current context, the highly contested political situation was unique to the period and the major source of transformation. Meanwhile, recruitment remains as, and arguably more, important to the development of the reserves as it did in previous eras. While the exact context may change, the numbers game has constrained the ability to transform the reserves and hence shape its future use and structure, and that of the army.

Perhaps more importantly, the current system of placing units on five-year rotations and tiered readiness cycles has political benefits. By locking units into a deployment, recovery and training cycle, 'which is how people defend against the civil service, the treasury and officials'[104] it becomes increasingly difficult to reduce the army any further without seriously threatening the coherence of the post-Fordist rotational system. Army2020, and within it FR20, therefore provides the kernel for which to expand on in a national emergency, and an insurance policy for the organisation against further defence cuts. Herein lies its originality. It is not only a solution to economic and strategic pressures that are similar to those of the past, but also a buffer against future politician's desire, and ability, to further reduce the size of the army. Such a political view of British military capability is perhaps even more applicable to the reserves, with its attempt to integrate formed sub-units into the army's readiness cycle. At considerable political and economic cost, FR20 has all but guaranteed the Army Reserve's organisational survival. These costs have included further damage to that identified in British civil–military relations during Blair's premiership, and also, as Chapter 6 discusses, forced politicisation of the army at the grass-roots level.[105] Nevertheless, FR20's underlying political goal, the reinvigoration of the reserves, has arguably already been met. But it remains to be seen if it will deliver the military capability originally envisaged. This fact may highlight the most important reason for the current transformation.

---

104 Interview, Lamb.
105 Bailey et al., *British Generals*.

Chapter 5

FR20: Delivering Capability?

The last three chapters discussed the historical evolution, the intra-party-political origins of, and the organisational concepts and processes underpinning, FR20. I showed how this post-Fordist approach is significantly different to how the British Army, and by extension the TA, practised logistics in the past. As a result of outsourcing logistics to the Army Reserve, its logistics units' missions are now more demanding than in the past. In order to understand capability, culture and cohesion in these units, it is crucially important to understand this changed nature of their organising principles and missions. The next three chapters examine the impact of FR20 on reserve logistics sub-units. As will be remembered, it is at the sub-unit level that FR20 originally envisaged the most profound changes to Army Reserve capability and deployability. Similarly, one of the greatest areas of risk for FR20 was deemed to be the reserve logistics component, as this required significant organisational changes as a result of the policy. This chapter discusses how the changes outlined in FR20, and the organisational frictions these have created, have impacted reserve logistics sub-units. In particular, these sub-units' experiences of what I term the 'hard' capability-related impacts of FR20 are examined. These are the interrelated issues of recruitment, equipment and training. Combined, these factors will ultimately determine whether sub-units can deliver the capability required of them under FR20. Throughout this chapter, how organisational transformation has been shaped by the post-Fordist approach to military logistics is analysed. The conclusion argues that while some sub-units may prove able to provide the required capability, most are unlikely to do so on schedule, and will be unable to provide an enduring capability for some time to come.

To understand the capability FR20 requires logistics sub-units to deliver, it is important to first outline these sub-units' specific transformations as a result of the policy. The outsourcing of regular army logistics capability to the reserves as directed in Army2020 and FR20 resulted in a number of major structural changes to the reserve component, including the disbandment of some logistics and infantry units, the creation of new logistics units and sub-units, the re-location of others, and crucially, the re-roling of some from one trade to another. As such, for some REME and RLC reservists, FR20 represented not only a complete change in the nature of their military specialist trade (trade training), but also a change of the location in which this was usually conducted. Within the RLC reserves, transport units bore the brunt of these changes. The REME also experienced major structural changes as the old system of Light Armoured Detachments (LAD) – a small team of specialist mechanics attached to other units – was replaced by a new centralised system based around the REME reserve battalion. Simultaneously, the rationalisation of the British defence estate saw the closure of reserve centres and the centralisation and co-location of units in larger bases.

It is also important to be clear about the terms used below. The British military definition of capability is the 'combination of equipment, manpower, and training to provide an effect or output.'[1] This essentially physical definition is supported by less-tangible moral group attributes such as cohesion, and discipline, which are generated and maintained through sub-unit ethos and leadership to maintain professional standards.[2] Intrinsically related to capability is effectiveness: the 'degree to which something is successful in producing a desired result.'[3] In sub-units, physical and moral capabilities therefore determine effectiveness. However, effectiveness is often role specific; the output on which it is judged can vary considerably between sub-units with different functions. Meanwhile, the ability of sub-units to be both capable and effective enough to execute assigned missions on operations is referred to as readiness.

The British Army uses a defined set of standards for both the capability and readiness for all regular and reserve sub-units. Reserve sub-units undergoing major transition as a result of FR20 were given two dates by which they had to deliver a certain level of capability. While these dates varied by unit, broadly speaking sub-units had to reach Initial Operating Capability (IOC) about 18–24 months after transition, while all units were to meet Full Operational Capability (FOC) in a similar time frame after IOC. While the exact sub-unit capability requirements for IOC and FOC are restricted, for a unit to be assessed as FOC it must be fully manned, qualified, and trained to the requisite standard. The FR20 schedule holds that most units should be at FOC by April 2019. Additional demands come from the readiness cycle, which is separate to the FR20 capability schedule. Across British defence a system is used which assesses units' readiness in terms of manpower, equipment and collective training.[4] The readiness cycle is used to bring units up to readiness for potential deployment, and can consist of boosts to manpower, equipment and intensified collective training. Reflecting their difficulties in deploying fully trained and qualified personnel, reserve sub-units' requirement to be at readiness is usually much lower than their established strength. For example, a 70-strong reserve sub-unit could be required to deploy a section of eight personnel – all qualified in the

1   Interview 17.
2   Ministry of Defence (2014) *Joint Doctrine Publication 0-01* (JDP 0-01), 5th Edition, Swindon: Development, Concepts and Doctrine Centre, 25.
3   Available at https://en.oxforddictionaries.com/definition/effectiveness, retrieved 21 September 2016.
4   National Audit Office (2005) *Assessing and Reporting Military Readiness*.

rank they hold – to support a regular unit on the first roule of a deployment. While the sub-unit could probably fill this requirement while only at IOC, its ability to sustain this requirement over subsequent roules would be affected if it was not at FOC. Understanding FR20's capability, effectiveness and readiness requirement is therefore central to answering the question of how the policy has impacted the selected sub-units.

Although there had been consultations and the drip-feeding of information before FR20 was formally unveiled in July 2013, all the selected sub-units had begun to be affected by the policy at the grass-roots level by the summer of 2014. This was when sub-units were informed of the date by which they had to deliver the IOC expected of their newly formed or re-roled sub-units. The IOC date varied from 2016–2017 between sub-units, and it was earlier than the FOC date at which these units are to be fully manned and trained. While the demand on each sub-unit varied in terms of the number of fully trained reservists they were required to deliver and at what stage of the tiered deployment cycle (organised into ten, six-month roules) this must be done, as a general indicator FR20 required these logistics sub-units of about 70 personnel to deploy 8–12 members to each roule. However, some sub-units may rotate this requirement with others in their regiment depending on the requirement after a certain date. Interestingly, none of the selected sub-units were expected to deploy as a fully formed unit, but to contribute a much smaller number of troops. Combined with other sub-units, a reserve RLC regiment would aim to deploy about a platoon's worth of reservists, while the REME requirement was greater, with a Field Company required. Clearly then, in the case of the reserve logistics component, reservists will also be deployed in smaller groups than the sub-unit. This was also found to be the case in other infantry regiments and taken together, this evidence supports the previous evidence on how FR20's original goal of deploying formed reserve sub-units has been adjusted as the policy developed due to organisational resistance.

Apart from their capability requirement, the nature of organisational transformation within the selected sub-units also varied considerably. For example, 165 Port and Maritime Regiment – responsible for the specialist loading and unloading of ship-based supplies in ports and onto beachheads – was expanded to become the largest reserve regiment in the RLC.[5] To fulfil this new capability, new squadrons were added to its order of battle; 232 Squadron (Sqn) – which had been part of 155 Wessex Transport Regiment and was manned by soldiers with advanced heavy vehicle driving qualifications – was directed to re-role to the port and maritime trade. This latter role required a completely different skill set to load, pilot and unload the Mexeflote sea-landing raft. However, it would remain in its Bodmin, Cornwall base. Meanwhile, 142 Vehicle Squadron, which, as part of 166 Supply Regiment was a nationally-recruited driving and maintenance unit based in

---

5   Interview, senior RLC officer, 25 March 2015.

Grantham, Lincolnshire, kept its trade but was moved to Banbury, Oxfordshire to become part of 165 Port and Maritime Regiment. 142 Squadron also incorporated a large number of former highly skilled Royal Signals reservists who had served near Banbury but whose unit had been disbanded as a result of FR20. Further north, 157 Transport Regiment was expanded, with the raising of 398 Transport Squadron in Queensferry, Wales. Simultaneously, Wrexham-based A Company of 3 Royal Welsh Regiment were told they were to be disbanded, leaving its soldiers the choice of a long commute to remain as infanteers, or re-roling with the new 398 Squadron much closer to home. Meanwhile, as part of the establishment of the new 105 Battalion REME reserve, one of its four new sub-units, 160 Field Company, was established at existing premises in Bridgend, while another, 130 Field Company, was established in Taunton. Both will centralise previous LADs in these regions, but 130 Fd Coy faces a much tougher task in reaching its established strength as there were fewer LADs in its catchment area, thereby limiting the number of already trained specialists it can draw on. With all LADs incorporated into new battalions, the re-structuring of REME reserve units was therefore slightly different to those in the RLC. Complicating matters, individual reservists with long distances to travel to these new sub-units frequently opted to join a differently traded unit nearby. As such, the scope of the transformation within these reserve logistics sub-units was often profound and represented some of the greatest organisational challenges posed by both FR20 and Army2020. Indeed, the army high command's awareness of the 'high risk' nature of this reserve logistics transformation is precisely why it is so worth of study here.[6] Given the different organisational changes experienced by different logistics sub-units, it is unsurprising that the data revealed varying experiences in the scope, nature and impact of FR20. However, a number of significant themes emerged.

FR20 Change Management

One of the first group interview questions asked respondents about how the organisational transformation of their sub-units had been managed. It is important to note here the initially high levels of uncertainty concerning the future of many logistics sub-units as FR20 planning took place during 2012–13. Many sub-units were aware that they were being considered for re-organisation but were not certain of the date by which this was to occur, nor its exact extent. Given the scale of the changes happening within the wider army as a result of Army2020, and the knock-on effect this had in terms of determining the required reserve capability, numerous interviews with sub-unit commanders indicated that they were only informed of their final transition plan in December 2013.[7] This uncertainty was reflected in

6   Interview 12.
7   Interview 3; Interview 5.

soldiers' experiences of how change was managed in their sub-units. For example, in one RLC squadron, the collective response was:

> Moderator (Mod): How do you feel the changeover was managed in terms of the creation of this unit?
>
> Respondent (R) 1: I can't fault it. I think it worked really well.
>
> R2: I agree.
>
> R5: We all knew the end goal. I personally thought it was all alright.
>
> R2: The final result was good.[8]

A REME sub-unit reported a similar experience:

> Mod: It went smoothly?
>
> R1: Yes.
>
> R3: I thought we were definitely kept in the loop. And in fairness to [the OC], he kept us up to date regularly, even the things he wasn't sure on, he was coming back to us and telling us. So yes, I was pretty informed ... We knew what was coming, didn't we?
>
> R1: Yes, it was managed really well.[9]

However, even in sub-units that had positive experiences, some individuals who had joined from different parent units felt they had 'got lost in the wash' due to a lack of information being provided by their chain of command.[10] Other reservists reported much more negative experiences:

> R1: Very poorly, it was done.
>
> R2: We all found out on Facebook.
>
> R3: I was told in an email ...
>
> R1: ... I think it was poorly conducted from the [original unit] side.
>
> Mod: At what level?
>
> R1: All.
>
> R3: I think it came from further up than regiment and battalion. I think it went up higher.[11]

---

8   Interview 1.
9   Interview 7.
10 Interview 1.
11 Interview 5.

It is therefore evident that some parent units and newly formed sub-units were better than others at keeping their soldiers abreast of developments.

Discussions with the one-star officer responsible for delivering the logistics capability outlined in both Army2020 and FR20 indicated his belief that leadership would be an important explanatory factor in determining sub-unit experiences of transformation.[12] This was supported by the interviews. This discussion in one sub-unit is instructive:

R1: [Our boss] was a part of the FR20 team so he was real pro.

R2: It [successful transition] was down to [the OC's] ... enthusiasm.

R3: He was very good in that respect.

R1: I wouldn't say just enthusiasm.

R4: [His desire for an] MBE.

R1: No. He wanted to make it work.

R1: Both for himself and for his blokes, I think.

R3: And he has worked. This squadron has worked.

R4: He's put the effort in.

R1: A squadron twenty miles down the road with very few new people in it has not worked. It's down to the person ... it's the personalities who run the squadron.

R1: ... He doesn't want failure.

R5: None of us in here want failure.

R3: And I think to answer your question, it was probably mismanaged by people higher, more senior to him and he's managed to keep the flak from us.[13]

The comments above underscore reservists' perceptions of the central role of leadership and personality in not only implementing FR20 at the sub-unit level, but also in protecting lower ranks from organisational friction caused by top-down transformation. The comments on the MBE are also instructive as they highlight the perceived relationship between mid-level commander's support for organisational transformation and the benefits this will have for their careers. Similarly, soldiers' recognition of potential individual reward as providing a motive for commander performance – and the rejection of this motive by

---

12 Interview, Brigadier Mitch Mitchell, CD CSS, Andover, 29 April 2014.
13 Interview 2.

other senior ranks – highlights the potential friction for commanders between delivering top-down transformation effectively and managing the longer-term interests of soldiers. Another interview revealed the possible source of this theme in the sub-unit and how readily junior ranks are willing to support leaders with their best interests at heart:

> R1: Our boss stood out in front of everyone and said, 'I'm going to forego my MBE if I get the things that you guys want', and at that point we were like, 'Sound.' It's a bit of a joke here, though, that he wants his MBE.[14]

Other sub-units reported similar levels of satisfaction with their commanders' management of change, and officers who 'jumped the gun' or 'got ahead of the game' by enacting transition as soon as it became clear what was required were frequently praised.[15] In terms of the transformation literature, mid-level leadership was therefore crucial to the perceived success of organisational change. Indeed, the role of these commanders may be more important in the reserves' case than in the regular army, due their part-time nature which means that these units have fewer points of contact with the senior regular leadership that instigated the transformation. This also raises questions about both the extent of pre-transformation consultation with these units by the Army and the FR20 plan's subsequent sensitivity to the realities of organisational change in them.

While most sub-units recorded relatively positive experiences of leaders managing change, there were a number of negative responses. Most of these were clustered around one sub-unit in particular that had been disbanded. Most of these reservists later joined the RLC or REME. For example:

> R1:   The CO never spoke to us and the RSM never spoke to us ... [They never said], 'This is the decision we've made and this is why we came to this decision that you will be disbanded.' So that was very poorly done.
>
> R2: We still don't know why we've been disbanded now.
>
> R5: ... I personally felt quite let down.
>
> R6: Yes. I agree.
>
> R5: After spending ten years with them.
>
> R7: ... The CO was just, like, 'It's closing. Deal with it ...'[16]

---

14 Interview 1.

15 Interviews 1, 5, 7.

16 Interview 5.

These experiences were repeated by senior ranks from the same unit:

> R1: So, you know, these guys have served 20-odd years as a [specialism] and then to be told, in one fell swoop, 'You will no longer be [specialism]. However, what you can do is you can go to the RLC.'[17]

Supporting the central role of leadership, these negative experiences of transition were largely blamed on individual commanders, with the lack of information and a sense of betrayal evident. There was a palpable sense in the junior ranks' discussion that their original sub-unit had been disbanded due to regimental politics and the wider, politically-imposed nature of FR20. The above quotes therefore indicate a perception that some commanders were relatively powerless in resisting organisational change imposed by the 'higher-ups', be they the military chain of command, or politicians. Similarly, there was recognition and resignation amongst these reservists that higher command's preferred option was for individuals from their disbanded sub-unit to join an under-recruited, newly formed logistics sub-unit – rather than continue their specialism in their original unit – in order to make FR20 a success.

In terms of how top-down military innovation was actually implemented at the tactical level, 232 Transport Squadron were unique amongst the selected sub-units as the chain of command directed which unit and trade they were to become as a result of FR20. Interestingly, soldiers in this unit did not appear disappointed by the fact that they were directed to re-role, while others in the same regiment were given a choice of new unit. Indeed, individuals in most of the other sub-units were given the choice of joining at least three other units which varied by trade. In the case of the disbanding or re-roling sub-units, such as A Coy, 3 Royal Welsh, or 142 Sqn, 166 Supply Regiment, this meant that these units hosted different events in which diverse units from the Army Air Corps, Intelligence Corps, REME, and RLC all pitched to attract transferees. Most respondents who experienced this were impressed that they were given a choice rather than being simply directed to join a new unit. For example, in 142 Sqn:

> (Mod): Were you guys happy with the way that was done, instead of it being directed, you had a degree of choice?
>
> All: Yes.
>
> R2: They put on showcases, which were quality.
>
> R3: It was amazing.
>
> R2: The Army Air Corps brought in helicopters and everything.

---

17 Interview 6.

R1: I felt like a school kid again.

R3: You went round all these different units going, 'These are the helicopters. This is a tank'.

R2: You're never going to drive this![18]

Humour aside, reservists originally from 3 Royal Welsh were visited by four units with different specialisms, while a reservist in another sub-unit responded:

The RLC gave a better presentation to people to come across. They sold it better. That's why so many of us came over.[19]

Similarly, some soldiers in 160 Field Coy REME were pleased that on the disbandment of their various LADs they were given a choice between joining the RLC, or remaining in the REME but moving to a centralised location. Of the selected sub-units, there was a trend that a core majority of transferees came from one unit. These were complemented by individuals from an array of other units who usually chose their new unit due to its proximity. Nevertheless, highlighting the varied experiences of individual reservists within these sub-units, and how many sub-units have incorporated individuals with different specialist skills keen to keep their location or trade, a medic reported:

R1: Yes, it was managed really well. We obviously got told we were transferring across.

Mod: Offered or told?

R1: Told. It was a little bit different for us ... We just changed the brigade ... There was no change.[20]

Thus, while most sub-units were offered a choice of new unit, this was accompanied by a sense of lack of real choice in some cases, and simple top-down direction in others.

Given that FR20 was a top-down imposed transformation, that most disbanding sub-units were offered a choice of future unit is highly interesting. Primarily it indicates the chain of command's awareness of the importance that location and trade have on reserve service, and in particular soldier retention. Simply designating sub-units to new locations or trades and expecting the majority of their strength to accept this was viewed as risking reservist retention. As such, from the outset, the transformation was undertaken with the goal of keeping as many

---

18 Interview 1.
19 Interview 5.
20 Interview 7.

reservists as possible satisfied, and hence retained, during the re-organisation. That such a quasi-market approach to future service was adopted, and that other units were then so keen to pitch to potential transferees, highlights the increasing realisation amongst commanders that unit strength – reliant on recruitment and retention – would become a core measure against which they would be judged as FR20 progressed. However, the significance of the availability of choice is deeper than simply the desire for well-manned sub-units. It also points to fundamental difference between how the reserves were reorganised under FR20 compared to the regular army under Army2020. For regular units, far less attention was paid to the impact that re-roling, or a change of location, could potentially have on retention. Directed re-organisation is more common in the regulars, and when compared to FR20, highlights the greater emphasis on choice given to part-time volunteers who can leave service at any time compared to their full-time counterparts. Thus, the element of choice in the FR20 re-organisation highlights not only the need to recruit and retain reservists from the outset, but also the variance between how regular and reserve change management was effected due to the different nature of their service.

Main Effort: Recruitment

The expansion of the reserves is a central tenet of FR20, and as discussed in Chapter 4, due to arguments between the army's senior leadership and politicians about how quickly this could happen – and subsequent recruiting problems – reserve recruitment became both a politicised and controversial issue. It also became the benchmark by which the media judged FR20's progress. It is therefore highly worthy of examination at the sub-unit level. During the numerous visits to army headquarters, regimental headquarters and squadron lines to arrange and conduct individual and group interviews, it was possible to read the concept of operations slides that are customary for commanding officers to display in their units along with Part 1 Orders. These slides provide an interesting insight into their respective units as they succinctly contain the commander's mission statement for the unit and the scheme of manoeuvre for how this mission will be achieved. Central to the basic concept of operations is the identification of the main effort, which sets the unit priority for the next year, or for the commander's time in charge (usually two years). What was highly interesting viewing a number of these slides across commands and trades during 2014–16 was that, without fail, the main effort in each unit was recruitment. Crucially, this indicated that most of the activity the respective sub-units were conducting was related to recruiting. Furthermore, on these visits, usually within two minutes of meeting the officer commanding – and indeed other ranks – they would mention their squadron's strength and recruitment. Indeed, as a former regular, I was often quickly asked if I wanted to join the sub-unit. This evidence gives some context as to how pervasive the 'numbers game' mentioned in the last chapter had become at the sub-unit-level.[21]

FR20's central emphasis on recruitment and its major impact on squadron activity after transition were evident in the group interviews. Indeed, the perception of recruitment's primary, indeed defining, importance to FR20 was repeated frequently:

R1: I don't believe there's any change ...

R2: The biggest thing's been a big push on recruitment.[22]

R1: I don't think there's a lot of difference between this and the re-org in '96. The only difference is the actual emphasis on recruiting.[23]

Another senior rank in the same unit concurred:

R2: we are carrying on as we did when we got re-org'ed the last time. There's not enough emphasis on trade and actually bringing us up to fulfil that role ... the emphasis is on getting numbers through the door.[24]

This focus on recruiting raised a number of related issues. Perhaps most importantly, as the quote above indicates, all ranks across numerous sub-units supported the assertion that with recruitment as the main effort, the balance between it and individual and collective training had been upset:

R1: Mainly we're doing recruiting and nothing else.[25]

R1: We haven't really done any of that, [training with regulars] really. We've had one or two little weekends.

R3: I think it's because our main aim is recruiting still.[26]

R1: All my time is spent helping them through their recruit process. I haven't got time then to be going to do my training.[27]

R1: I think without a doubt it's been recruitment, recruitment, recruitment ... recruitment is number one priority for most reserve units; however, it's closely followed by retention and trade training.[28]

---

21 Interviews 1, 2, 5, 6, 7, 8, 9.
22 Interview 10.
23 Interview 2.
24 Interview 2.
25 Interview 1.
26 Interview 8.
27 Interview 5.
28 Interview 4.

R1: To get us to our 30,000 before 2020, which is where the emphasis is, as opposed to making us a better unit to actually support the people that we are supposed to support.

R2: Yeah.

R3: It's changed slightly this year, hasn't it?

R3: I mean, 2015, all of a sudden, they've realised, if they don't put some trade on then nobody is going to go anywhere [i.e. their careers will not progress].[29]

As the last quote highlights, soldiers in some sub-units did perceive the lack of training had begun to be addressed, but it is clear that FR20's drive for numbers, initially at least, created friction between recruitment and training to deliver capability. Clearly, senior command knew that it would take time to expand the reserves, and that collective training to confirm capability would necessarily follow this expansion. However, the problems recruitment has posed in terms of time and effort highlight an organisational paradox inherent in FR20 between expanding sub-units and delivering quality training that ultimately retains soldiers. The prevalence of this issue was confirmed in follow-up interviews in 2016. The quotes from the below interview are particularly indicative as they are taken from one of the most positive sub-units in 2015, which was fully manned a year later. Here, the emphasis on recruitment was still viewed as negatively impacting trade training and ultimately, retention:

R1: It's been nice to have a group and start having more people coming through. So that has been a big positive ...

R2: [but] there's no retention.

R3: And no training aids for the boys ... There's nothing here really is there?

R4: Weapons is a nightmare ...

R3: There's just no way of teaching anything.

R4: This is across the board.

R1: Just the bare bones of what the unit should have and nothing more.[30]

Thus, the lack of trade training and its impact on retention was still widely perceived as being caused by a lack of basic equipment in reserve centres. As another soldier stated: 'Our recruiting target is over the 100 per cent mark, but keeping them interested is another thing.'[31]

---

29 Interview 2.
30 Interview 8.
31 Ibid.

The emphasis on recruitment has created other organisational problems. Numerous reservists cited the saturation of recruitment teams from different reserve units competing for the same recruits in their region as an example of an uncoordinated wider approach to recruitment. The drive to recruit in ethnically-diverse areas was also seen as unsuccessful in those sub-units which had attempted to do so.[32] The quote below emphasises the lack of planning and resourcing of recruitment activity at the sub-unit level:

> We've still got Army Reserve recruitment teams who are not trained, not equipped, setting up army recruitment stands with white vans. Are we recruiting for Ford or Vauxhall, because that's what we're selling? We're not marketing it correctly. We're not providing the training for the recruiting teams and we're not being selective enough in those who are recruiting for us. We're not looking to attract the main target audience. Standing in a high street on a Wednesday afternoon with a white van and a gazebo is not going to recruit people for the Army Reserve.[33]

This lack of recruitment resources in recently raised logistics sub-units is particularly interesting because, given the high risk nature of their transformation, their centrality to FR20, and the difficulty reserve logistics units usually have recruiting compared to the combat arms, it might have been expected that they would have received extra support. This has clearly not been the case. Crucially, the lack of resourcing of recruitment activity was widely perceived as resulting from the outsourcing and centralisation of recruitment to Capita. As one senior NCO in another sub-unit explained: 'I wanted to get £50 to put up our details on the boards at the [local rugby team]. The loops we had to go through with Capita … it simply wasn't worth it.'[34] Similarly, reservists in other sub-units were aware of the initial problems and delays the Capita contract had caused in reserve recruiting, indicating a collective wariness about its centralisation in general.[35]

Moreover, the perception within most sub-units was that the emphasis on the quantity of recruits had come at the expense of quality. An RLC senior rank summarised this attitude best: 'Yes. This unit's conversion rate is good. It's about quantity. Really it should be quality.'[36] Within this theme, a number of issues were identified, the first of which concerned new recruits. Some reservists believed that new recruits were less suitable for military life than in the past

---

32 Interviews 1, 2, 3, 6.
33 Interviews 6, 14.
34 Interview 14.
35 Interviews 10, 1.
36 Interview 10.

due to increasingly sedentary lifestyles,[37] others that recruits were too young.[38] Another squadron noticed that recruits were younger, more likely to be female, and that they had provided some 'brilliant', committed new members.[39] Others were more outright in their criticism, specifically focusing on how physical standards had been dropped in order to increase recruitment:

R3: There are people [in the recruitment process] that shouldn't be. There's absolutely no way.

R1: The big one that I've noticed is because it's such a numbers game, the amount of fucking dross that we're getting through the door, that they suddenly say, 'Right, that's recruitment, you've got to bend over backwards for them', you just think, 'Why am I wasting my fucking time?' ... Some of them can't even do press-ups or sit-ups.

R3: ... You're probably talking less than 50 per cent that we reckon will actually be able to go through.

R1: ... should be able to turn around and say, 'Look, come back in three months' time when you're fit.'

R1: Back then [before FR20] ... you had the authority to do that, whereas now ... I've spoken to all high ranks saying, 'Look, it's not going to happen.' 'Put it in a letter.' So I put it in a letter and nothing happened.[40]

However, perhaps reflective of the more specialised skills required of their trade – and the higher aptitude scores required during soldier selection – the attitude was noticeably different in REME squadrons.

R1: I don't think the quality of the recruit is any different to how it's always been ... as long as they are eager and they fit certain criteria [mechanical aptitude], then it doesn't matter ...[41]

R1: We do go for quality. We know we cannot just take anybody ... You've got to take people on with some form of mechanical electrical knowledge. We've tried it before with people who were non-mechanic ...

R2: And it shows.[42]

---

37 Interview 1.
38 Interview 2.
39 Interview 10.
40 Interview 1.
41 Interview 7.
42 Interview 8.

Overall, therefore, RLC reserve squadrons appeared to be more concerned about the quality of recruit they were attracting, while REME companies appeared to be insulated from this because mechanical aptitude was a recognised requirement for successful service. Although there was recognition that 'some good ones have come through'[43] soldiers across the REME and RLC stated that some of these had been lost due to problems with the outsourcing of the reserve recruitment process, which some still perceived as too slow despite recent attempts to expedite it. Another reason for the failure to retain suitable recruits was the successive nature of modular training which can force a recruit who misses an important weekend to wait up to six months for another, thereby delaying their individual advancement.[44]

The commitment and the ability of new recruits – and indeed existing reservists – to meet the minimum required professional standard for reserve service was related to this quantity versus quality debate. This minimum standard was perceived to be the completion of the required Military Annual Training Tests (MATTs) as directed by the army. The inability of some reservists to compete these was a particular source of ire, further indicating the dichotomy between FR20's emphasis on quantity and delivering reservists of the expected professional standard:

> R1: The reserves are treated with kid gloves. 'We don't want to upset them because they might leave.' I say bollocks to that … if they're old and they can't do the fucking job, get rid of them.

> R2: Exactly. That is a massive smack in the face … if you can't pass your MATTs and you're not fit then why should you get your £2000 [bounty] a year like everyone else does that puts the effort in?

> R3: It all goes back to numbers.

> R4: If you did that, you'd get a few leaving. You need people on the books. It doesn't matter if they're fat.[45]

The quote above highlights how the numbers game has affected the retention of reservists who are failing to meet the required standards. This issue of the physical fitness of recruits and reservists, and in particular logisticians whose primary role is not combat, is likely to continue as fitness standards for all reservists (including previously low readiness logistics sub-units) were standardised with the regulars in 2012. As the comment above indicates, this has created friction between the professional standards of the regulars and retention in the reserves.[46] Related to

---

43   Interview 2.
44   Interviews 2, 8.
45   Interview 1.
46   Interview 14.

the collective desire for committed, professional members, another commonly expressed opinion concerned the quality of ex-regular transferees attracted by the £10,000 bonus – paid over three years – for joining the reserves. This incentive, introduced in 2014 after it was clear former regulars were not transferring in the expected numbers, has been very successful in increasing ex-regular recruitment. However, it is not without controversy:

> The ex-regulars are coming in because they are thinking: "Wicked, we are getting £10,000 over a period of [three] years" … but they are not turning up for training. They've come in to do their bare minimum … we really want the regulars because we need to learn from them. But they see it as an easy bus ticket … and then what are we getting out of it?[47]

This reservist again highlights the expectations of professionalism and commitment she expects of those in the reserve. Furthermore, the belief that some ex-regulars were not committed to the reserves due to the monetary incentives on offer was repeated in a number of other interviews.[48] While discussed in detail in the next chapter, it is worth noting here that the fact that some reservists felt aggrieved at the commitment of some regulars is itself indicative of the growing professionalism of the reserves.

Striking the right balance between recruiting committed reservists and offering the right monetary benefits to attract and retain them has become an increasingly important issue given the controversy over recruitment quantity and quality. Since 2013 the government has spent tens of millions of pounds on recruitment campaigns that have struggled to gain traction. It has therefore introduced substantial joining bonuses to both ex-regulars and new recruits, who are offered £300 on attestation and a further £2,000 after completing their first years' training commitment. Following previous research,[49] a quantitative examination of reserve RLC and REME soldiers' reasons for joining was undertaken to complement this study. The data revealed that a greater percentage (76 per cent) of the statistically-significant sample joined for institutional, intrinsic reasons – such as to be challenged or to serve their country – compared to occupational, extrinsic reasons, such as for monetary benefits and occupational development (63 per cent).[50] Crucially, institutionally-

---

47 Interviews 7, 17.
48 Interview 2.
49 Moskos, C. (1977) 'From Institution to Occupation, Trends in Military Service' *Armed Forces and Society* 4(1), 41–50; Griffith, J. (2007) 'Institutional Motives for Serving in the U.S. Army National Guard: Implications for Recruitment, Retention, and Readiness', *Armed Forces and Society* 34(2); Moskos, C. and Wood, F. (eds) (1988) 'Introduction' in *The Military: More than Just a Job?* New York: Pergamon-Brassey.
50 Bury, P. (2017) 'Recruitment and Retention in British Army Logistics Units', *Armed Forces and Society*, 43(4).

motivated soldiers were found to have longer career intentions and were more committed to reserve service.[51] This data was supported by the interviews:

> R1: That was the difference. They wanted to come and find you. Now we're going into town going, 'You can walk and you can breathe. You'll do ... Sign this bit of paper, son.'[52]

> R1: You want someone who really wants to do it – like I really wanted to do it. If they're half-hearted, they will fall out. And there's not many people that I think that really want to do it. You have to think ... I always look for the next challenge. That's what I do. That's how I am.[53]

> R1 The whole point of this [FR] 2020 is about saving money. Take a couple of steps backwards. How much is it really costing with all these financial incentives? Are they really going to save money?

> R2: After three years, are they going to stay in?

> R1: Exactly my point being is, these 'retention [bonuses]' ... are not really. They're sweeteners, not retention, because, once that money runs out, people are going to [leave].[54]

These reservists clearly believed institutionally- and intrinsically-motivated recruits are required in the long term. However, the institutional-occupational distinction is not mutually exclusive. One reservist remarked how she 'joined for the experience but now it's about the money',[55] while another senior rank argued:

> When you tell them what they can get out of it: licences, this, that and the other, it's a massive eye-opener. But it's the way we sell it ... they don't go deep enough in actually saying: 'Look. You get paid for this. You get this, you get that.' And people say, 'Oh, money shouldn't come into it.' Of course it does. This is Cornwall. It does count as an income to a lot of the guys. The majority of people in Cornwall are on low income, minimum wage. Well, if you can top that up being in the Army Reserve then well and good.[56]

Monetary and career development benefits clearly play a major role in recruiting and retaining reservists. The question is therefore about where the right balance lies between institutional and occupational recruitment models. The fact that some

---

51   Ibid., 19.
52   Interview 1.
53   Interview 10.
54   Interview 8.
55   Field notes, 8 May 2015.
56   Interview 11.

regulars' motivations were perceived as circumspect due to the bonuses on offer, and that compared to the past, recent recruitment campaigns have highlighted the (increased) material benefits of service, raises interesting questions about the long-term commitment of reservists recruited by campaigns that stress occupational benefits. Indeed, there is an awareness amongst senior officers of the potential risks of the occupational recruitment model.

Another major sub-theme was the degree to which soldiers perceived the recruitment drive as resulting from politics. This was related to a lack of confidence in the Army Reserve's ability to recruit to its target strength.

> R4: I don't think they're doing enough to make it a success. I think that people have actually used this for political gain and ... they're not really supporting. They're not investing.

> R1: ... The politicians have got it wrong ... To attract young people it's a real challenge. They [young people] don't understand it.

> R2: They [politicians] put a lot of weight on all the regulars that they kicked out on joining the TA and it hasn't happened.

> R3: No military boss is going to cut their army. It's got to be the politicians, hasn't it?[57]

> R1: What they wanted was all the regulars made redundant to join the TA. That's what they wanted ... but that's not going to happen.

> R2: Politicians are causing all this. They call all the shots ... that's what it comes down to.[58]

The quote above is also noteworthy for an awareness of the failure of regulars who had been made redundant to join the reserves in the expected numbers due to the initial lack of incentives on offer, further supporting previous arguments about the ad hoc evolution of FR20. Meanwhile, the political theme was repeated frequently across ranks and sub-units:

> Everyone in the army knows this plan is politically driven.[59]

> The 30,000 in seven years is pie in the sky.[60]

> Yeah, we recruit to our targets ... But will we reach it [full sub-unit strength] by [2019]? No.[61]

---

57 Interview 6.
58 Interview 5.
59 Interview 19.
60 Interview 2.
61 Interview 11.

Senior ranks were also aware that total sub-unit membership did not reflect real, trained strengths:

R1: It's not accurate, either.

Mod: What isn't? The recruiting?

R1: The numbers. We've got 72 people on the books, because we've got 72 members, but of that I would say there's only 40 per cent that turn up, and they'll get their certificate of competence at the end of each year. So, in actual fact, from a squadron of 70, there's only 40 people.

R2: That is your true strength.[62]

Although significant given FR20's emphasis on numbers, this finding is not as controversial as perhaps it first appears. In the 1980s Walker recorded that TA turnout varied between 40–70 per cent of total strength and that most units had a 'hard core' of about 40 per cent.[63] The interviews revealed that this situation suited some sub-units as it gave them a greater allocation of Man Training Days (MTDs; the metric by which funds are allocated to units and which ultimately determine training) to be utilised by core members.[64] Another sub-unit noted problems in retaining these new recruits.

R1: You have loads of people coming in, but then very few stay.

R2: We have as many go as we do come in the door.

R1: … We could have a battalion here. The amount of people I've seen in seven years come through these doors, and actually got into the regiment and uniform, and even some of them have gone up to the basic training and finished. And they're just … gone.[65]

While the same problems were not replicated across the sample, overall there was a general trend that RLC squadrons were less optimistic about their ability to recruit to full strength by 2019, despite the effort these sub-units had expended to date. REME squadrons were generally more positive about their ability to get to full strength. While overall experiences were mixed, it is clear that major issues remain concerning recruit quality, commitment and retention in some of these sub-units.

However, the most controversial finding concerned the reporting of sub-units' strengths. With a 40 per cent core of regular attendance, in most reserve units

---

62   Interview 6.
63   Walker, *Reserve Forces*, 107–108.
64   Interview 11.
65   Interview 10.

another 30 per cent attend once a month and the remainder rarely. Usually, once a reservist has not attended for six months, they will be struck off the sub-unit books and the discharge process instigated. However, this study revealed numerous cases where sub-units had been instructed by higher command to keep personnel who had not attended for over one – and even two years – in order to show their strengths were rising and hence support the narrative that the Army Reserve is growing as a result of the recruitment drive. One example from the fieldwork is particularly instructive. Arriving at a sub-unit, I was shown its personnel roster displayed on the wall by a concerned senior soldier. Of the approximately 50 reservists on it, one third were highlighted in yellow as not having attended a parade for a year. The soldier explained: 'Some of these haven't attended in two years or more. We've been told not to shit-can anyone 'cause it looks bad on stats ... we've got to keep these on our books to make our stats look good ... If I went to the papers with this they'd have a field day.'[66] The soldier explained that he understood the direction to effectively 'cook the books' had come down from higher command in the past year, but that it would be reversed soon. It was not the aim of this research to uncover who issued this direction, nor when, but the general finding was supported by numerous other sources. As these soldiers explained in separate interviews:

The other thing, especially the senior ranks, because they know more and they've been around more, is the deceit that is put on the news channels that 'we've recruited this many' and we know all these numbers aren't true. It's creative accounting at its very best, because we are here in a squadron of 110, 120, and there's how many people tonight? 12? 10 per cent.[67]

R2: The problem is it's that bit of paper there and it's 'how many numbers are on that bit of paper in that book?'

R1: It's just so wrong.

R2: That's what it boils down to. Somebody will open that and go, 'Oh, your books are looking good.'[68]

R6: From what I see of it, there are a lot of paper soldiers on the books in the reserves.

R2: Yes.

R6: Half the people who are actually on the books, which the government figures are on target, they don't exist.

---

66 Interview 20.
67 Interview 11.
68 Interview 2.

R1: But … there's not a commanding officer who's in charge of a reservist unit who's going to go, 'Ah, not turned up. Strike off the books,' or, 'Not suitable. Strike off the books.'[69]

Similarly, questioned on recruitment, an officer in another sub-unit offered the following analysis:

It's quantity over quality. And to be honest it's forcing us to play politics. We say our strength is higher than it really is and the chain of command pass that up. Of course they know it too … it's forcing us to be political really.[70]

A recent newspaper report corroborated this data with evidence from other units, with 'senior military sources' also stating that reservists who fail their fitness or weapons handling MATTs were not being discharged, supporting the evidence presented above.[71]

These are significant findings for a number of reasons. Most obviously, they indicate that some units have been directed to report inflated strengths in the full knowledge that in reality up to a third of this strength are not active members. It also indicates that the chain of command, in areas where it has not directed this to happen, is still complicit in it. However, perhaps most worryingly, it indicates the degree to which sub-units and, perhaps the chain of command, have been forced to support what is an inherently political plan. In doing so, it raises major questions about transparency in Army Reserve strength figures. Perhaps more worryingly, as the officer above hinted at, this practice is blurring the traditional – and legal – line between politics and service in the armed forces, thereby making some senior soldiers and officers uncomfortable with the manner in which they are forced to support FR20.

Equipment, Training and the Limits of Post-Fordism

FR20 pledged to invest £1.2 billion (although £500 million of this was in cancelled cuts)[72] in Army Reserve equipment and training over ten years, and in order to understand whether sub-units were becoming more capable, reservists were asked whether they perceived increased levels of equipment and training in their sub-units as a result of the reforms. Strachan and King have both shown how training is crucial to understanding cohesion and effective performance in the

---

69   Interview 7.
70   Interview 21.
71   *The Daily Express*, 3 January 2016, 'British military reservists hugely undermanned, say army whistleblowers'.
72   Personal correspondence, Professor Vince Connelly, July 2018.

infantry.[73] While basic specialist infantry drills are relatively easy to conduct with soldiers and personal weapons – and RLC and REME reservists do learn basic infantry skills – the availability of equipment, and especially vehicles, on which to train is particularly important in these sub-units as without them realistic trade training is difficult to conduct. Thus, the availability of vehicles to regularly train with is central to retaining the specialist skills and knowledge required for these sub-units to collectively deliver the capability outlined in FR20.

One frequently occurring theme was the lack of vehicles and equipment available at sub-unit locations:

> R1: From a training perspective, on a Tuesday night we don't have a chance to do anything for vehicles. Skills training, yes you can do that. You could probably do with more kit like pistols.[74]

> R1: We get the basic uniforms, but we haven't got the main kit like trucks and weapons and radios. We still haven't got that.

> R2: We're still a bit Dad's Army.

> R3: We've got nothing. When we first started, we were doing infantry lessons, skills, but with no weapons. The equipment is there. We just can't get hold of it full-time.

> R2: … We've already lost a year's training [because we] still [had] no trucks.[75]

> R2: I manage the equipment here. At the moment, we hold barely any equipment within the squadron. Is that a regimental fault? I don't know. So, as to equipment, no, we beg steal and borrow from the other squadrons. We've got minimal equipment just to keep the squadron afloat …[76]

While there was an acknowledgement that equipment would become available as transition progressed – and there was clear evidence of this during the research[77] – it was clear that this had had an impact on numerous sub-units.[78] However, soldiers in a REME field company did note the better availability of equipment since FR20 and the positive impact this had had on training, indicating that experiences varied in this regard:

> The operational kit has started rolling through now, which never used to happen.[79]

---

73 King, *The Combat Soldier*; Strachan, 'Training, Morale and Modern War'.
74 Interview 1.
75 Interview 5.
76 Interview 6.
77 Field notes, 21 February 2016.
78 Interview 14.
79 Interview 8.

Meanwhile, following the rationalisation of the defence estate, the lack of infrastructure in reserve centres was another theme that indicated problems with the synchronisation of transition. This appeared in numerous interviews:

> … the infrastructure wasn't in place. We're now a year down the line. We've got, on our account at the moment, 40 weapons only. We're still waiting for the equipment to come … Things like that [haven't] been handled well. I think any unit going into a location, the infrastructure should be in place. The equipment table should be in place. Then, the troops come in. To get the troops in first, and not have the infrastructure in place, is not great.[80]

Other soldiers succinctly commented:

> The Army's come up with this master plan. The stuff should have been there before we came over.[81]

> The fundamental thing that was wrong when we first started was that by the time we started we had no infrastructure whatsoever.[82]

Other units reported an initial lack of showering facilities that hindered physical training, while a lack of offices was also a problem.[83] The fact that a number of centres did not have Defence Intranet terminals was also problematic as it made conducting personal administration tasks more difficult.[84] However, other units saw the beneficial aspects of estate rationalisation, and the centralisation of units this had caused:

> R1: [It] was very hard for command and control, unit cohesion, and everything like that. It was very difficult. Now, it's under one unit …

> Mod: It's centralised?

> R1: Yes. It is so much better. You're recruiting for one location. You've got control in one location. So, that's been a big bonus.[85]

There was also a large degree of acceptance, and patience, that these infrastructure issues were an initial result of the re-organisation and that they would eventually be resolved.[86]

---

80 Interview 3.
81 Interview 5.
82 Interview 8.
83 Interviews 5, 6, 13, 14.
84 Interviews 5, 6, 14.
85 Interview 3.
86 Interviews 5, 6, 7, 13.

However, it was interesting to note reservists' perceptions that many problems with equipment availability were the result of centralising kit while outsourcing its management to civilian firms. This was apparent across RLC and REME sub-units and occurred without respondents being asked directly about the effects of centralisation or post-Fordism in general. For example:

R1: Apparently the British Army doesn't own any low loaders [heavy transporters] now. It sold them all.

R2: Yeah. [Outsourced] Contract.

R3: Mr. Witham [civilian military equipment sales firm] will now sell them back to us at twice the cost and get another MBE for it.[87]

R1: The one thing that seems to be a little bit confusing is they're downsizing the regulars, expanding the reserves, but they're downsizing the size of the [vehicle] fleet ... If people are going to go down the [centralised] whole fleet management side, forget it. They've tried that before. It's a bag of shit ... I've been here 30 plus years and I've seen all fleet management tried many a time before.

R2: Plus whole fleet management works for regs because – this is the regs and the reserves thing again – because [if] you're deploying this weekend, you need five vehicles. That's five drivers, plus a driver to drive those drivers to get those vehicles.[88]

The frictions created for reservists by centralised equipment stores and outsourcing was repeated in other sub-units:

R1: For us to do a trade training weekend, we've got to go all the way ... to Marchwood ... They've got a hell of a hike to get up there.

R1: That's a lot of time travelling.

R3: I've also heard that Marchwood was being sold.

R2: They've sold it, and so the military are only going to get limited hours using it.

R3: And we don't know how that's going to affect the equipment and kit there, whether we're not going to get it, or whether it's going to come to Plymouth.[89]

---

87 Interview 1.
88 Interview 8.
89 Interview 11.

The reference to the sale of Marchwood Sea Mounting Centre to civilian port management firm Solent Gateway in October 2015 is particularly interesting in terms of the impact of the post-Fordist approach. The port is under-used in peacetime and as a result the government expects to generate revenues by allowing Solent Gateway to use it for 215 days per year. The remaining 150 days' use is shared between the regular and reserve port and maritime RLC units who train and unload vessels there. While the contract was in part negotiated by the regular RLC officer responsible for the port at present, the 150-day limit has raised concerns over availability of the port for reservists, given they do not usually train during the week when the regular units would be using their allocation of days. As such, the outsourcing of the port was designed more with the regular, rather than the reserve component, in mind.

The interviews also revealed problems with the plan to outsource maintenance tasks from the regulars to the newly centralised reserve REME battalions as part of the integrated whole force concept:

> R1: ... they're trying to run it [equipment maintenance] like a regular battalion ... We just haven't got the manpower. It cannot be done.

> R2: Production's limited during the week because we're only there for a couple of hours Wednesday nights. So you can't undertake big tasks.

> R2: They sold it to us originally that this is all aimed at keeping competencies up for the tradesmen ... But I think that we're doing less now than we were when we were [as LADs] with the other regiments.

> All: I'd agree with that.[90]

Clearly, centralisation, outsourcing and the attempt to better integrate the reserves following the network approach indicate that the adoption of post-Fordist principles is having major impacts on these sub-units' ability to train and, within the REME at least, also on the regular's expectations of them to share the maintenance burden.

In sub-units that reported a lack of equipment and/or infrastructure, this was seen as a major obstacle to delivering routine collective trade training:

> Mod: How do you think the move, then, and the transition has affected training?

> R1: Dramatically.

> R2: Yes, because the vehicles ... the biggest seller for an RLC transport unit is vehicles ... They want to do training in the Man SV [truck]... However, what [reservists] require is the vehicles for them to drive. If they haven't got their vehicle to drive, then they're going to start to walk [leave the unit].

---

90 Interview 8.

R3: … We're not just planning for training at the minute. We're planning to collect kit so we can train.[91]

This last quote underscores the problems caused by centralised equipment stores for reserve sub-units.

R1: Due to the lack of kit, Tuesday nights are becoming repetitive. We are running out of subjects that we can cover. There's only so many times, without practical training, that you can deliver theoretical training …

R4: The infrastructure problems will have an effect on recruiting, because a [young recruit] will walk in and have a little look at this building. 'Do I want to be part of this organisation?' Yes, we can mask things up … but we're meant to be attracting people, not taking people to a throwback from the '80s. They will not go into a working environment that is below standard.[92]

While these senior NCOs' comments, and those in other sub-units, underscore the relationship between equipment, infrastructure and training, soldiers in the same unit also highlighted the importance of being able to collectively and routinely train in their specialist trade in order to prevent to the loss of skills learned on individual qualification courses.

R1: It's alright smashing out the courses, but unless you all get together, and not just over a weekend, [collective competency will not improve].

R3: You can't really develop if you haven't got the kit. So it's all going to come back to kit.[93]

As such, the lack of equipment was seen as a crucial impediment to conducting collective training, and it combined with the emphasis on recruitment activity to further threaten retention. While this was especially apparent in logistics units, it should be noted that, in general, reservists personal kit has been updated as a result of FR20 to bring it in line with that of the regulars. As discussed below, while the success of FR20 in RLC and REME sub-units will ultimately rest on their ability to provide enough individuals trained to the required level of trade specialisation, individuals across ranks and sub-units expressed that the lack of routine collective trade training was causing skill-fade and low confidence.[94] Due to FR20's tiered readiness system, as long as individual reservists have the right course qualifications, this could be addressed during mobilisation before operations,

---

91 Interview 6.
92 Interview 6.
93 Interview 5.
94 Interviews 1, 6, 14.

but it is clear that the longer sub-units are left without equipment the more likely it is to negatively impact both morale and collective readiness.

Despite the problems a lack of equipment was causing, a number of sub-units did report improved training opportunities in the wake of FR20:

> Now we are part of a battalion, our training is more REME orientated, its more focused on us and what we need to do.[95]

> I think we're more capable of doing the job, because we've got more resources now, more manpower ...[96]

> One thing I've noticed is you get more opportunities to do more live taskings ... it used to be up to a few years ago ... 'No, you can't go.' Whereas now, for these port tasking groups, they can't get enough guys to go on it so it ends up with 50 per cent regulars, 50 per cent reservists.[97]

Given that FR20 aimed to increase the capability of these sub-units, these quotes indicate that progress is being made with training despite the difficulties outlined above, and there was also evidence that access to equipment was improving as the research progressed.[98]

A related issue concerned training with the regulars. FR20 stated that reserve units would be paired with their regular counterparts to achieve better integration under the Whole Force concept. Interview groups were therefore asked their thoughts on pairing and whether they had conducted frequent training with the regulars since FR20 was introduced. Complementary quantitative data revealed that within the wider RLC/REME reserve population some of the most negative attitudes to reserve service were related to the amount of training undertaken with regulars,[99] but responses were more varied across the sub-units that were interviewed. Some units that had trained with regulars were satisfied with the quality of training and how they had been treated by their paired units;[100] others reported few opportunities to train but blamed this mainly on transition;[101] while some sub-units had negative experiences of working with regular units.[102] While this variance is to be expected and likely reflects both reserve sub-units' differing transition schedules and the different command climates within regular units, increasing regular-reserve collective training opportunities was identified by the army as a key method of

---

95   Interview 5.
96   Interview 6.
97   Interview 1.
98   Field notes, 21 February 2016.
99   Bury, 'Recruitment and Retention in British Army Reserve Logistics Units'.
100 Interviews 1, 10.
101 Interview 7.
102 Interview 5.

delivering both better integration and reserve retention in the wake of quantitative data collection. The evidence presented here complements this data in this regard and suggests that more could be done on the issue.

Despite the importance of collective training to deliver capability – both in reserve centres and with the regulars – the fundamental determinant of whether sub-units will successfully re-trade on schedule rests on their ability to train, or re-train, individuals to the standard required of their rank in their new specialism. As a result, the ability of sub-units to get reservists on specialist trade courses is central to the success of their transition. Without individuals trained to the required standard, these sub-units simply cannot provide the required number of trained personnel to deliver the increased capability expected under FR20. As such, courses are critical to understanding the trajectory of transition in sub-units that have re-traded; they are the building blocks of sub-unit capability. Although there was some evidence of problems with the availability of RLC courses in particular,[103] across the sub-units there was a general acknowledgement that FR20 had dramatically increased course availability to members.[104] For example:

> I think the support's a lot better. I think personal development's a lot better …[105]

> The PSIs [Permanent Staff Instructors] have done fantastic getting people on courses.[106]

> We're mainly smashing through getting people the right licences. As soon as they've got the right licences, we're getting them on the trade courses to get them to that level. Again, it takes time.[107]

> [The availability of courses is] a good thing that should be publicised more, I think. If you've got the time to put in to it, you're going to near enough get what you want out of it.[108]

Numerous other sub-units were equally complementary about their regular PSI's efforts. However, at an organisational level, given that individual proficiency provides the foundation on which sub-unit capability rests, the better availability of courses post-FR20 is to be expected.

Nevertheless, the interviews and fieldwork revealed major frictions associated with re-training entire sub-units that indicate serious, and perhaps fundamental, weaknesses in the FR20 plan to transform these units. Of these issues, the most

---

103 Field notes, 8 May 2015; Interview 1.
104 Interviews 1, 5, 7, 14.
105 Interview 5.
106 Interview 6.
107 Interview 6.
108 Interview 12.

serious concerned the tiered progression of specialist trade courses that are tied to the rank structure. For example, to qualify as a sergeant in a port unit, a reservist would be expected to have completed their class B1 course indicating that they can safely operate numerous heavy vehicles used to load and unload ships in a number of conditions, and direct their safe use as well. To get this qualification a reservist would have first had to complete his class B3 qualification course, gain two years' experience in trade and complete complementary modules, then complete his B2 (perhaps as a corporal), wait another two years and complete further modules, before finally completing his B1 and qualifying as a port sergeant. In the current system, this would take a reservist a minimum of six years if completed as quickly as possible. Re-roling sub-units has therefore created a situation whereby higher ranks – those vital NCOs with deep trade experience upon whom sub-units rely to deliver training and capability – must now learn a new trade. They are effectively in a rank without the relevant professional skills demanded of their trade. While the training problem was less severe for junior soldiers who simply follow the new trade progression, given the timelines involved in gaining trade experience and relevant qualifications, and the pressures on reservists' time, the depth and scale of transformation in sub-units that have been directed to re-role cannot be overstated.

Overall, the difficult reality of re-trading reservist logisticians was starkly apparent in sub-units that had re-roled. The interviews below provide context for the impact these changes have had on both individual and collective capability:

R1: You're doing your B3, your B2, your B1. You're probably looking at about five or six years.

R2: Whereas we're all a little bit screwed, those that have come across from the Signals. I was a Class 2 CS [Communications Specialist] op[erator] and now I've had to go all the way back to the very beginning to sit my B3 course. I can't get promoted to [corporal] until I've got B2 upgrade and I've got my HGV licence. I've got a job. How on earth can I ever fit this in?

R1: Just to do the track licence that's four weeks you've got to take off. Impossible.

R2: ... They've worked it out that it would take us a couple of years to go to the next level.

R3: That's providing you've done all the courses.

R2: So we're pretty fucked.

R4: Which ... when they sold it to us, that wasn't actually explained.

R1: The best position to be in, coming across, would be as a private.

R6: I've come across [as a private] … So I did two weeks [last year]. This year I will do two weeks. Then I'll be in the same boat as everyone else.[109]

We've got a lot of guys that have come in that are not role-specific, so they need to be beefed up to the training levels required. The soldiers that are transferred across [in trade], that's a given, but we've also got soldiers coming in through the door. Now, when we make the offer, we say, 'Come and join us, you'll do your training. You will then go and get your driving licence. You'll be converted across.' So we've got quite a big backlog of training that needs to be done. We can't organise sub-unit training during training times because we haven't got the kit on-hand. There's a certain amount of hours that people need to do behind wheels to prove competency, to make sure that they're road legal. I really don't think the higher echelons understand the training bill for a reservist to get in, in his or her 27 days, because that is what they're committed to.[110]

Apart from the importance of the sheer amount of time required for reservists to re-trade, the loss of experience and knowledge – and an acknowledgement that this cannot be gained simply on courses – were also frequently highlighted:

R1: For me, as a full screw [corporal] RLC now, you've lost nine years' experience. So when somebody asks you something, you honestly don't know.

R1: The courses might be [there] but the experience won't be …[111]

By the book, we've got people that are already trained, but they don't have the knowledge. They've done the courses. Fine. The knowledge and experience takes time.[112]

This need for further training and deep experience is enshrined in the qualification process itself, but in order to expedite this process it appeared that reservists were allowed to attend courses without gaining the required amount of experience. As alluded to, such a practice is not without risk and, paradoxically, may actually increase the gap in real-term capability between the regular and reserves that FR20 was designed to decrease. Indeed, it provides further evidence of the ad hoc nature of the policy and the level of self-harm it caused by being implemented by the army without sufficient sensitivity to the organisational realities of re-trading reservists and rushed through to save money.

---

109 Interview 1.
110 Interview 6.
111 Interview 5.
112 Interview 6.

As highlighted above, the need to re-train soldiers, and in particular NCOs, negatively affected their careers, causing numerous qualified personnel to quit as the re-organisation got underway. While a two-year dispensation was quickly given to NCOs in new trades that didn't have the relevant qualifications in the wake of re-organisation,[113] re-trading was seen as potentially detrimental to their promotion:

R1: In this squadron, if you look around the different ranks, there is only possibly three people who are qualified to move to [Staff Sergeant], one person who is qualified to move to Warrant Officer, and that's because they came across with their trades intact. But the guys who have come across are in an unfortunate position where they miss out.

R2: In limbo.

R3: Career failed.

R4: Well, they are not career failed because everyone has to go through same re-training progress so it's not a career fail because you've got five or six seniors who all have to go through the same training process.[114]

The diverging opinion expressed by the last respondent is interesting as it suggests that as re-trading has affected all senior ranks uniformly, the competition for promotion between them has therefore remained the same. While true, this view does not appreciate that this practice effectively delays these ranks' promotion, pay and seniority compared to if they had remained in trade. This was explained by one senior rank: 'I am positive about the transition … [But] one of the biggest knocks was promotion.'[115] Similar views were expressed in other units that had re-traded.[116]

It is also interesting note how the need to re-trade was viewed by some senior ranks as indicative of the limits of the Whole Force concept in general, and how the reserves remained separate from the army in particular. For example:

R1: If it's supposed to be one army, you've achieved a rank. And with that rank, you've got additional management experience. Management experience is supposed to cut across the army and it doesn't seem like that has been taken into account. So we are not one army. We are still the individual corps.

R2: I think you've hit it on the head. We're not one army. We never will be.

R3: That's right.

---

113 Interview 22.
114 Interview 2.
115 Interview 11.
116 Interview 6.

R1: ... This new strapline of one army doesn't work because it's not ... It can't be. Because we are on two different levels, and they need to stop pushing that because [it] winds everyone up.[117]

Similarly: 'They say it's one army but it's got two separate pay scales.'[118]

What is clear from the above comments is that FR20 has created extensive organisational friction in these sub-units concerning the realities of re-trading and its impact on careers. During the research, it became clear that higher command was aware of the scale of challenge re-trading already qualified NCOs posed. Although much uncertainty remained, there were suggestions that trade-qualified NCOs would keep their old trade and be promoted without the required trade specific courses.[119] In this scenario, junior soldiers would continue to train for the new trade, while seniors would not be required to do so. Sub-units would therefore lack experience of their trade at the senior rank level, and without help from other units this would clearly affect collective training. Either way, the juxtaposition between re-trading to deliver the required operational capability and correctly managing career progression cannot be easily solved. While on the one hand the re-trading issue underscores the considerable organisational friction that FR20 has generated between the delivery of collective capability and the correct career management of reservists, on the other it indicates that transformation in these sub-units is almost certain to take longer than FR20 originally envisaged. Again, given the history of the previous periods of reserve reform this is hardly surprising.

Can These Sub-units Deliver?

What do the above findings tell us about the impact of FR20 on these sub-units to date? Clearly, sub-units have had different experiences, and generally it was noticeable that those that had experienced the least amount of organisational change appeared to be the most content about FR20. It is also important to stress that by the completion of research in summer 2016, FR20 was three years into its five-year lifespan. The interviews also revealed that many soldiers had not passed final judgement on FR20 yet:

In a couple of years' time, we/ll know whether the plan was a good one or not, its probably too early [now].[120]

---

117 Interview 2.
118 Interview 5.
119 Interview 22.
120 Interview 2.

There's lots of changes going on, but it's a bit early to say whether they're for the better or for the worse. [So far] I think they're better, from a REME point of view.[121]

Interestingly, this last position had noticeably changed a year later in the same unit:

Mod: Do you think in your experience FR20 has been positive? What's the general consensus?

R6: Mixed.

R2: Mixed.

R6: It's good for the fact that we bring in loads of people in and get the numbers up, but I think it's bad for retention.

R2: Lots of stuff that should be happening is now slipping, like the kit and the training aids.[122]

As a result, although this chapter utilised longitudinal data, the long-term nature of FR20 limits the ability of it to provide a definitive answer about the policy's impact on the sub-units in question. Nevertheless, a general sense of what FR20 has achieved to date, the rate of change, and its future trajectory, can be ascertained.

Recruitment, equipment and training are some of the most important 'hard' factors that will determine if these sub-units can deliver the operational capability required by FR20. To ascertain whether those in the selected sub-units thought this possible given FR20's impact to date, reservists were asked whether they believed FR20 would ultimately prove successful. Again, there were mixed responses, with the following opinion from an NCO in a sub-unit that had not re-roled reflective of others commonly aired: 'On paper they'll succeed, whether you'll see a huge difference on the ground is another matter.'[123] Meanwhile, it was widely acknowledged that re-traded sub-units had become ineffective as a result of FR20, and it would take years to re-develop effectiveness in a new trade. As one officer remarked:

So for six years, if we're lucky, we will be ineffective or we'll need to borrow the senior guys [from other sub-units in the same regiment]. How does that help unit cohesion, integrity etc? It doesn't. Like I said, we're never going to have a ganger, a sergeant, because it's six years, [but] we can probably end up providing, with help from [other sub-units]. We could then put all our

---

121 Interview 8.
122 Interview 12.
123 Interview 12.

resources together and provide ... But as a squadron, independent, with the manpower we've got, and the trade [issue] ... We can't do it.[124]

This belief that their sub-unit would not ultimately prove able to provide the FR20 capability requirement was repeated in other sub-units, but it was also challenged:

R1: We've still got a long way to go.

R2: Will we be able to [provide] 40 personnel? No.

R3: I disagree. I personally think by 2017 we could have 30 odd people ready to deploy.

R2: They would be deployable. They have done the courses. They'll have the fitness, but ...

R1: It's the knowledge.

R2: Experience.

R4: You need to be competent before you can be deployed, because otherwise you're putting people at-risk. They will go at-risk. They will then have a really bad time ...

R3: To gain the experience, they need the kit and equipment. My argument is yes I would say, by 2017, we will have between 30-odd people ready to deploy, as long as we have the equipment to put them on the road and ready for deployment.[125]

No, we're not ready. We've only been formed six months and next year they're expecting us to be at high readiness state. No. Definitely not.[126]

R1: We will be in the same position we are in two-and-a-half years.

R2: I don't think we will.

R3: If things don't change, then yes, we will.

R4: We could do it [deploy a section] now.

R1: ... In two-and-a-half years, we'll still be able to deploy a section.

R2: But it would be top-heavy.

R3: ... It would be top-heavy because of our trade.[127]

---

124 Interview 11.
125 Interview 6.
126 Interview 8.
127 Interview 2.

This identification of the 'top-heavy' nature of sections refers to the deployment of senior ranks in junior positions in order to provide capability. As such, it reflects the perceived inability of sub-units to fill each section with the required trade skills relevant to rank. Critically, it indicates that, because these sub-units would be forced to deploy more of their senior ranks in the first roules of a deployment, their capacity to continue to deliver the same capability in subsequent roules would be severely limited. It therefore indicates potentially major problems in the tiered readiness system within which these sub-units are meant to provide capability. As an officer elucidated:

> [This sub-unit must provide] a half section for roule one and then a full section for roule two, three and four, right up to eight. Now, [there] is an interesting mathematical conundrum at this point ... because I'm supposed to have five years' worth of [deployable reservists], and I've only got four. So my last two roules, either I cover them with the first two or ... we can train specialists in that length of time, so that my ninth and tenth roules are covered by people who are not yet through the door ... [But] I would actually say it [the tiered readiness cycle] is a wise idea. It's something that can work. We cannot do it on existing philosophies, but I think we've done quite well in adjusting philosophies to make things happen. Am I capable now of delivering a section? Probably, yes. Am I able to deliver a section, plus a section into the barrel for six months' time? Probably, yes. But beyond that, I'm going to work quite hard and I know that. It's something that I've known for over a year. And we're making great progress. But, as regards Full Operation Capability, I don't think we'll ever get it, because it means that everybody has to be trained all the time.[128]

Interestingly, officers and soldiers in other sub-units also identified this weakness in the roule system, but were less positive about their ability to bridge the capability gaps they identified in the sixth year of a deployment.[129] Such a finding raises serious questions as to the long-term sustainability of the reserves logistics components contribution to the Army2020 deployment system, and, more generally, of the Whole Force concept.

To gain an accurate picture as possible of the impact of FR20 to date, items on this topic were also included in the surveys, which were distributed to a wider sample that was less weighted toward units which had undergone organisational transformation. The results were more positive than those from the group interviews, but still some of the least positive opinions expressed in the surveys overall. Perhaps the most important baseline statistic concerning the impact of FR20 to date is recorded in the item concerning respondents' optimism that the

128 Interview 3.
129 Field notes, 21 February 2016.

reforms will increase their sub-unit's capability. On average 54 per cent said they 'Agreed' or 'Strongly Agreed' with the proposition, 30 per cent 'Couldn't Say', and 16 per cent said they 'Disagreed or Strongly Disagreed'. In terms of perceptions of sub-units becoming better at their job as a result of FR20, the jury is still out, as 43 per cent stated they 'Couldn't Say', while 38 per cent 'Agreed' or 'Strongly Agreed'. Only about 32 per cent agreed that there had been better equipment as a result of FR20, and this item recorded some of the least positive results in the entire survey, suggesting that the flow of better equipment into sub-units has been moderate at best. In terms of FR20 delivering better integration with the regulars during training, there were slightly more positive results, with 33 per cent stating that they agreed this had occurred in their sub-unit. However, 24 per cent of both cap badges recorded 'Disagree' or 'Strongly Disagree' responses. Supported by other findings above, and previous research, this indicates that increasing training with the regulars has not been fully met yet.

Despite the varying experiences of transformation and differing opinion on its likelihood of success, this chapter has shown that FR20 has already had major impacts on the sub-units examined. Many of these were foreseen at the outset of FR20 implementation; some have been unintended. Overall, how the policy was implemented has created considerable organisational frictions in these sub-units, with many long-term issues still outstanding. FR20's emphasis on recruitment has come at the expense of training, which, paradoxically, can pose a threat to retention. The numbers game has also raised questions about the quality of recruit and whether they will remain committed to service with these sub-units in the longer-term. Perhaps most importantly, the need to recruit to target for political reasons has caused accounting practices to develop that lack transparency and have ultimately forced commanders to play politics.

The impact of certain post-Fordist logistics structures and management processes on equipment, infrastructure and even recruiting in these sub-units has also been evident. The centralisation of logistics equipment, especially vehicles, has reduced the amount of training time for many sub-units on a number of their key equipment platforms, while also increasing the complexity of, and human resources needed to, conduct trade training. Similarly, the evidence suggests that within this sample, the Whole Force concept is yet to be fully implemented, with reservists concerned about the lack of availability of equipment and training with the regulars. Meanwhile, the centralisation of reserve sub-units in larger base locations has resulted in improved command and control in some sub-units, but also infrastructure and recruitment problems in others. The sale of Marchwood port is particularly instructive as it indicates the negative impact that rationalising military infrastructure can have on reserve units whose time is more limited. Similarly, the centralisation and outsourcing of the previously localised reserve recruitment process to a civilian firm has caused major impediments for the reserves. While some of these issues are being addressed, overall the evidence suggests that while the post-Fordist approach to logistics may deliver efficiencies for the regular army, as it has not been designed or implemented with the reserves in mind, it may create more problems than it solves. It is arguable that, given their part-time nature and more local dispersal, the post-Fordist approach is far less efficient and useful to reserve sub-units than it is to their regular counterparts.

Despite FR20's promises of investment, many sub-units are suffering from a lack of equipment and infrastructure, and this has negatively impacted training. While there was an acceptance in the sub-units that equipment and infrastructure issues are being addressed, that they existed was perceived as a result of poor management by higher command and the politically-imposed nature of FR20. This failure to resource these logistics sub-units from the outset has clearly impacted positivity about FR20 and is related to the ad hoc nature of FR20's development. Decisively, this chapter has also shown that in sub-units that have re-roled, re-trading has created a situation whereby some sub-units are likely to take six years or more for them to deliver even their IOC. While some sub-units may be able to deliver IOC on schedule, real transformation in many sub-units is likely to take longer than the April 2019 date by which FR20 is due to be completed. Similarly, even if reservists are pooled at the unit level in order to deliver the required capability for initial deployment roules, such are the organisational challenges created by FR20 in these sub-units that many of them will lack the capacity to sustain their contribution to later deployments, thereby jeopardising the tiered, rotational readiness structure, not only for reservists in these sub-units, but also in the regular units they are designed to support. And this is notwithstanding the longer-term issues for reserve sub-units identified within the rotational system itself. As a result, in the case of the sub-units examined here, it currently appears that many are unlikely to be able to deliver the full operational capability required under FR20 on schedule. While some, especially REME companies, will, and most are likely to meet demand of at least their first allocated roulements, given the scope of the organisational changes experienced, it appears that many will struggle to provide the enduring capability envisaged by FR20 – and upon which the army relies for Army2020 to function – for many years to come.

Chapter 6

FR20 and Cohesion

The last chapter presented qualitative evidence from sub-units on their ability to meet the capability requirements set by FR20. In order to understand if and how FR20 is affecting cohesion and readiness in reserve units, this chapter proceeds in two parts. Firstly, it uses the Standard Model approach to examine quantitative data from a wider sample of reserve logisticians to assess perceptions of cohesion and readiness, experiences of FR20 to date, and the impact of these experiences on cohesion over time. Secondly, drawing on revisionist cohesion scholars' methods and analyses, it utilises qualitative evidence to complement and deepen understanding of FR20's impact on reserve cohesion. In doing so, for the first time it incorporates two very different but complementary approaches to cohesion in a single study.

Two levels of analysis are used in the initial survey discussion. A statistically-significant sample of the RLC/REME population is utilised to illustrate individual perceptions in this group, while the sub-unit level is utilised as it provided more stable sample groups for longitudinal comparison. The aim was to produce statistically-significant findings reflective of the wider Army Reserve RLC and REME population. While the sample was statistically representative in the first tranche of data collection in 2015, in 2016 responses were much lower, and thereby it should be stressed that the longitudinal sub-unit comparisons, while internally highly statistically valid, are only indicative to, not representative of, the wider REME/RLC population. The findings concerning sub-unit changes over time therefore represent an initial sketch of regular and reserve cohesion, rather than a definitive conclusion. Importantly however, the results of the 2015 data are all representative of the wider population.

Standard Model Survey Results[1]

A useful classical definition of cohesion under the Standard Model has been provided by Guy Siebold: 'The level of unit cohesiveness is defined as the degree

---

1   Some of the evidence and text presented in the following sections originally appeared in Bury, P. (2018) 'Future Reserves 2020: Perceptions of Cohesion, Readiness and Transformation in the British Army Reserve', *Defence Studies*, 18(4), and is kindly reproduced with permission.

to which mechanisms of social control operant in a unit maintain a structured pattern of social relationships between unit members, individually and collectively, necessary to achieve the unit's purpose.'[2] Siebold originally identified three basic components of unit cohesion: horizontal, vertical, and organisational. Each component is conceived of having an affective (emotional or feeling, known as social cohesion) aspect and an instrumental (action or task, known as task cohesion) aspect. To assess these components the surveys asked soldiers to rate their perceptions of different aspects on a 5-point (very low, low, moderate, high, very high) scale. In terms of the results, the average score of the 2015 cohesion survey was 4/5 for each question, indicating relatively high perceptions across all the components of cohesion. This is an important baseline statistic as it indicates that levels of perceived cohesion amongst REME and RLC reservists are similar to those recorded in studies of both regulars and reservists.[3] This indicates that despite substantial organisational changes within some of the sub-units surveyed, these soldiers' perceptions of their sub-unit's cohesion remain relatively high. Indeed, they are comparable to those recorded in regular forces. Overall, these reservists' perceptions of cohesion were therefore positive in 2015. Similarly, perceptions of discipline and leaders setting values were relatively high.

Military readiness can be defined as the ability of military forces to fight and meet the demands of the national military strategy. At the sub-unit level, readiness refers to the unit's ability to carry out assigned missions. This study uses the definition of morale provided by Gal: 'A psychological state of mind, characterised by a sense of well-being based on confidence in the self and in primary groups.'[4] Overall, the results showed relatively high levels (3/5) of readiness and morale. About 33 per cent thought that their sub-unit's readiness was in the high categories, 48 per cent in moderate, and 17 per cent in the low categories. Nevertheless, the fact that the majority reported moderate over high readiness is noteworthy, especially when compared to the cohesion results. Interestingly, soldiers' perceptions of their individual readiness to fight if necessary was significantly higher (49 per cent in both cap badges responded 'High') than perceptions of sub-unit readiness. This included high levels of confidence in the sub-unit's major equipment systems (56 per cent), although of note is that on average 35 per cent said their confidence in this regard was 'Moderate'. In terms of individuals' confidence in their ability to do their job on operations given the correct pre-deployment training, 87 per cent

---

2   Siebold, G. (1999) 'The evolution of the measurement of cohesion', *Military Psychology*, 11(1), 18.

3   Siebold and Kelly, *The Development of the Platoon Cohesion Index*; Siebold, G. (1996) 'Small unit dynamics: Leadership, cohesion, motivation, and morale', in Phelps, R. and Farr, B. (eds), *Reserve component soldiers as peacekeepers*, Alexandria, VA: U.S. Army Research Institute for the Behavioral and Social Sciences.

4   Ingraham, L. and Manning, F. (1981) 'Cohesion: Who needs it, what is it, and how do we get it to them?' *Military Review*, 61(6).

reported 'High' or 'Very High' levels of confidence; 83 per cent reported similar levels of confidence in the ability of their sub-unit to perform on operations given sufficient pre-deployment training. Sixty-six per cent also stated that their sub-unit's morale was in the high categories. This is another important baseline statistic and coupled with the fact that only three per cent rated their sub-unit morale as 'Low' and none as 'Very Low', indicates high levels of sub-unit morale across the sample. This is especially positive given the organisational changes many of the sub-units have experienced due to FR20. Indeed, high levels of personal morale were also recorded (71 per cent in high categories), in stark contrast to comparable data on morale in the regular army.[5] Overall then, in the first tranche of data collection in 2015 the cohesion, readiness and morale scores were generally positive.

The results were similar in terms of experiences of working with the regulars, 55 per cent agreed that working with the regulars had increased their confidence in their individual skills. Slightly lower levels of agreement (46 per cent) with the regulars' impact on sub-unit competence were recorded, with higher levels in the 'Can't Say' category (44 per cent). Sixty-five per cent agreed that working with the regulars was a valuable experience. The results were less positive concerning reservists' confidence in FR20 delivering on its objectives. In 2015, 54 per cent said they agreed with the proposition, and 30 per cent 'Couldn't Say'. However, a major interest was to ascertain if sub-unit perceptions of cohesion, readiness and morale, and experiences FR20, were changing as FR20 progressed. In terms of the three sub-units selected for longitudinal comparison between 2015–2016, it is noteworthy that their average cohesion scores remained relatively stable, but that in one sub-unit this had dropped by about four per cent. There were no significant changes in these unit's readiness and morale scores either. Taken together, these statistics highlight that FR20 does not appear to be increasing cohesion, readiness, or morale over time. Conversely, the 2016 data indicated significant positive increases in the mean scores in units' attitudes to working with the regulars since 2015. This likely reflects more exposure to the regulars, and positive experiences during this increased exposure. As such, this data supports the qualitative data presented in the last chapter that FR20 is increasing reservists' exposure to the regulars. However, there was a decrease in confidence that FR20 would deliver increased sub-unit capability over the same time period, at similar levels of significance (-9.81 mean, Sig=.00, t= -8.22, df =32). Importantly, this contradicts the data presented for the wider sample in Chapter 5 and indicates that in the three sub-units examined longitudinally, confidence in FR20 has declined since 2015. When combined with the lack of significant data concerning growth in cohesion, readiness and morale, this supports my argument that FR20 is struggling to increase sub-unit capabilities but is making progress with integration with the regulars.

---

5   Ministry of Defence (2016) *Armed Forces Continuous Attitudes Survey 2016*, available at https://www.gov.uk/government/uploads/system/uploads/attachment_data/file/523875/ AFCAS_2016_Main_Report.pdf, retrieved 21 September 2016.

Finally, the survey results also revealed an interesting broader trend. The regular sub-units reported lower perceptions of Affective Bonding (social cohesion), and higher perceptions of Instrumental Bonding (task cohesion) than their reserve counterparts. While this would initially appear to indicate that there are lower bonds between regular soldiers, in fact when taken in tandem with the regulars' higher instrumental component scores, this actually may suggest the greater importance of task cohesion in the regulars. This supports the qualitative analysis below.

A Different View of Cohesion[6]

In *The Combat Soldier* – the most in-depth examination of professionalism's impact on combat forces to date – Anthony King argues that the intensive collective training associated with professionalisation has gone beyond the mere transformation of Western forces' effectiveness to fundamentally alter the nature of social relations between their soldiers.[7] Drawing on Max Weber's concept of status honour in uniting groups, and the threat posed to these groups by the heightened individualism of Emile Durkheim's anomie, King shows how both of these have imbued military practice with a moral force – a professional ethos – that unites military groups.[8] Thus, expanding on Huntington's identification of the importance of 'corporateness' amongst the professional officer class – in essence their shared commitment and sense of community – King argues that the enhanced emphasis on training and collective action in the professional infantry instils a common obligation to perform effectively, not just amongst officers, but combat soldiers as well. For King, 'professional comradeship' based on effective performance has replaced the classical sociological understanding of cohesion based on interpersonal bonds.[9] Professionalism has superseded love as the main source of cohesion in the infantry.

Although a later study by King and I examined how 'cold professionalism' influences, and is influenced by, the heightened social and emotional bonds of combat,[10] thereby partly reconciling the classical cohesion literature with that on professional militaries, the recent cohesion literature has focused exclusively on the battlefield performance of combat forces. In fact, with a few notable

---

6   Some of the evidence below originally appeared in Bury, P. (2017) 'The Changing Nature of Reserve Cohesion: A Study of Future Reserves 2020 and British Army Reserve Logistic Units', *Armed Forces and Society*, available at https://doi.org/10.1177/0095327X17728917.

7   King, *The Combat Soldier*, 339.

8   Ibid., 341.

9   Ibid., 350.

10 Bury, P. and King, A. (2015) 'A Profession of Love? Cohesion in a British Platoon in Afghanistan' in King, A. (ed.) *Frontline: Combat and Cohesion in the Twenty-First Century*, Oxford: Oxford University Press, 213.

exceptions, the majority of the cohesion literature to date has focused on infantry soldiers.[11] This leaves open the question as to whether the nature of cohesion in logistics units is the same as in the infantry. Moreover, to date, the focus has been on professional regular forces, leaving further uncertainty about how reserve forces – with less time to train intensively and therefore, theoretically at least, lower skill levels – generate and sustain their cohesion. Indeed, neither the wider reserve literature, nor that on the British reserves in particular, has conducted a detailed examination of reserve cohesion, nor the impact of professionalism upon it.

Both Kier and Farrell have discussed the importance of normative, cultural aspects in explaining military transformations, but neither sought to investigate how culture and cultural emulation manifest themselves at the micro-level. Both approaches were also top-down and concerned regular combat forces. Here, I discuss the distinctive nature of reserve logistics units in order to examine the FR20 transformation from the bottom-up. Firstly, I compare the nature of cohesion in logistics forces to the infantry, using field observations of the selected sub-units as an evidential base. Upon clarifying this issue, the central question of FR20's impact on cohesion and professionalism within reserve logistics sub-units is discussed from a qualitative perspective. The subjects include the persistence of social cohesion; the demise of the 'drinking club'; the rise of professionalism; and the unique nature of reserve discipline. I suggest that, following the last chapter, although many of these sub-units are unlikely to deliver the full capability required of FR20 on schedule, in many areas the transformation is slowly but profoundly changing the culture of, and social relations in, the Army Reserve.

Logistics and Reserve Cohesion

In order to assess the impact of FR20 on these sub-units' cohesion, first it is necessary to briefly examine the nature of logistics and reserve cohesion in general. King's work on the importance of intensive, repetitive, standardised training in explaining cohesion is also supported by those of Ben-Shalom et al. and Strachan. For these authors, cohesion relies not on interpersonal social bonds between soldiers, but on the effective performance of the military group. Both Ben-Shalom and King examine infantry platoon training in great detail, arguing that standardised words of command and individual battle drills, co-ordinated at the collective level, are central to explanations of successful combat performance. Indeed, King's work on infantry combat techniques illustrates the importance of minute, almost esoteric, movements such as the position of the thumb when

---

11   Zurcher, L. (1965) 'The Sailor Aboard Ship: A Study of Role Behaviour in a Total Institution', *Social Forces*, 43(3).

firing, and the need to 'bob' around corners to reduce angles in urban combat.[12] For him, such techniques are indicative of the resources, knowledge and time devoted to training in the professional infantry. Such minutiae are important because, when implemented correctly at the individual level, and crucially, co-ordinated at the collective level, they reduce the risk of fatalities. One of King's observations is therefore that the threat of death in combat as a result of the failure to execute drills correctly provides another powerful explanatory factor in explaining successful collective infantry performance.

Most logisticians, however, are not trained to the same standard in infantry techniques. Despite the need for infantry skills when operating in non-linear battlespaces where insurgents threaten supply convoys, logistics soldiers' primary role lies elsewhere. Unlike the infantry, where different units share standardised skills and training techniques, in logistics units the skills required of personnel are as varied as logistics functions themselves. In the British Army, these functions include transport and movement; port and maritime operations; explosive ordnance disposal; air dispatch; catering; cleaning; and post duties, amongst others. Meanwhile, REME craftsmen specialise in the repair and maintenance of numerous vehicles and airframes. There is therefore a vast difference in the skills required of logistics soldiers compared to infanteers. Moreover, there is a difference between logistics trades requiring high levels of co-ordinated collective action and others reliant on individual skills. This distinction is important in understanding the nature of cohesion in logistics units.

As its name suggests, 165 Port and Maritime Squadron RLC is specialised in two trades, the former being the loading and unloading of cargo in ports and beachheads, the latter referring to the delivery of cargo on seaborne rafts. The port trade requires operating a variety of heavy plant, including forklifts, trucks, bulldozers, and specialised 40-tonne cranes. While these are essentially individually executed tasks, often, the driver of the vehicle is assisted by a senior rank who directs operations through standardised hand signals and words of command, highlighting the applicability of King's arguments to explanations of performance in certain individual logistics trades. Such action also requires detailed practical knowledge of how to load and secure all types of cargo for movement by road or sea. The maritime trade's main operating platform is the Mexeflote raft. The skills involved to operate this are varied. Highly trained engineers are responsible for piloting the raft and repairing its engines, while another NCO is responsible for the safe loading of the Mexeflote and instructions to the engineer. Junior ranks are responsible for correctly securing the cargo using different lashings, and the safe landing of the Mexeflote. Thus, individually and collectively, the skills required in this logistics unit are vastly different to those of the infantry.

---

12 King, *The Combat Soldier*, 320.

The Persistence of Social Cohesion

The fact that some logistics trades are individual and others require collective action has major implications for how reserve logistics soldiers generate and sustain their cohesion. For King, the cohesive bonds in the professional infantry are formed by intensive training and commitment to their profession.[13] Following Leon Festinger, he also notes that the density of interactions amongst infantry soldiers is important; propinquity matters.[14] However, intensive training requires large amounts of time unavailable to part-time reservists. With less time to train and shorter qualifying courses, there is also a skills differential between most regulars and reservists. The part-time nature of reserve service also suggests that the density of interactions is less than the regulars. Thus, theoretically, in reserve logistics units requiring collective action, social cohesion may still be important. It would be expected to be even more so in individual trades. However, FR20 is attempting to professionalise the reserves through increased training in order to better integrate them with the regulars and prepare them for routine deployment. This creates an interesting problem in terms of understanding their cohesion. On the one hand, reserve logistics units may be more reliant on social bonds than the regulars. On the other, reserve professionalisation could be changing the nature of social relations in these units and hence their sources of cohesion.

One final discussion is needed here on the impact of the post-Fordist approach to logistics on the skills and cohesion of the logistics units examined. Although I have detailed that the term post-Fordism mainly refers to strategic and operational management processes and structural issues, there was some evidence that post-Fordist innovations will eventually have an impact on logistics skills at the tactical level. As discussed in Chapter 3, the most obvious example is that the nature of modern conflict and the adoption of the nodal FOB system reliant on CLPs has forced logisticians to hone their combat skills to a higher degree than in the past, resulting in a greater emphasis on military skills in logisticians' pre-deployment training programmes. Another example concerns the MJDI system. While it still has some teething issues to be resolved, when fully introduced and integrated with TAV-, soldiers reported that it should significantly quicken vessel unloading by alleviating the need to manually stock-check and process items before onward transportation. The introduction of MJDI was also acknowledged to begin a radical change in the structure of RLC support to combat units and hence changes in the career structures and training of RLC soldiers, who will now have fewer postings to combat units.[15] However, the need to attend MJDI specialist courses in order to operate the system has itself provided other opportunities in this regard, highlighting the ongoing specialisation within the 'core' logistics component. Outside of MJDI and other

---

13 King, *The Combat Soldier*, 374.
14 Ibid., 351.
15 Interview 23.

IT-enabled software systems, and the impact of centralisation and outsourcing on training discussed in Chapter 5, there was little evidence that post-Fordist logistics was changing skills at the tactical level in the units examined. Nevertheless, this will likely change as new technologies emerge to support the distributed logistics model; for example, a road transport unit could conceivably be re-roled to operate delivery balloons in response to the full automation of road vehicles. While such changes could force the source of certain units' cohesion to change, with more emphasis on individual rather than collective skills, overall the nature of cohesion in logistics units is unlikely in the short-medium term to differ from the individual/collective paradigm outlined above. Nevertheless, automation and robots may also ultimately render many logistics skills, and hence units, redundant, depending on task complexity and threat environment.

While Leonard Wong, and Bury and King, have shown how interpersonal bonds still contribute to effective performance in the professional infantry, there has been no detailed examination of cohesion in the reserves since professionalisation.[16] In his 1990 study, Walker briefly discussed cohesion in the TA, highlighting the importance of 'drill hall club' social cohesion. For Walker, this referred to the beers usually enjoyed in messes by reservists after weekday training at their sub-unit location, and he opined that the 'activities at the drill hall club are perhaps as important as the evening training itself.' Crucially, he noted that social cohesion built 'the regimental esprit and unit identification critical for sustaining not only combat units, but also volunteer reserve units in which cohesion is a precursor for encouraging men to turn out for training.' He also observed the importance of social events for generating cohesion and how reservists' 'social life begins to revolve around the unit'.[17] For Walker, interpersonal bonds remained central to understanding TA cohesion in the late 1980s. Of course, when Walker was writing, only social cohesion had been identified. Thus, the question remains if, as the AR professionalises, social cohesion remains as important as it did in the past?

In order to assess FR20's effect on reserve logistics cohesion, first its nature was considered. To do this, interviewees were directly asked about the value they placed on the social element of service. The social element was left for respondents to define and describe as they wished; some referred to time spent socialising in the bar, others the bonds between colleagues. However, across ranks, the unanimity and strength of response was notable:

The social element, I think, is important.[18]

That's a massive thing.[19]

---

16 Bury and King, 'A Profession of Love?'.
17 Walker, *Reserve Forces*, 102, 105–106.
18 Interview 3.
19 Interview 5.

It's important. It's important for … a bit of an army lifestyle, isn't it? You've got to have that.[20]

R1: It's key. It can't be all work and no social element, because it is a lifestyle. The guys do deserve to get rewarded and there needs to be a balanced work/ social environment.

R2: The guys are giving up their free time. They don't have to be here. So there's a point where you need to sit down and give them something back.[21]

It's massive.[22]

R1: Hugely important.[23]

Clearly, the social aspect of reserve service itself – that which generates and sustains interpersonal bonds – is deemed very important to logistics reservists. Primarily, it is viewed as an intrinsically vital part of these soldiers' service, but also as a reward for part-time volunteers' commitment to training and duty. Moreover, this social element, with its organised events and socialising in the bar after duties, was stated to be reflective of the regulars. As others elucidated:

R1: I joined the reserve after leaving the regs because I was missing the craic [fun] with the boys, the laughs, the banter …

R2: One of the best things about the regs is the social life, you know, your summer balls and your Christmas balls. Again, regular social events … seem to have carried over to the reserves as well.[24]

That's why I joined, really, [to] meet new people.[25]

For these reservists, social events are normalised by reference to regular practices. Moreover, this social element provided a primary joining motivation for these soldiers. Previous surveys revealed that 86 per cent of these reservists had joined to make new friends in the military, the second most cited reason for joining.[26] Overall, this data indicates that socialising together is very important to reservists' motivations for joining, and for generating and sustaining social solidarity once they have.

However, the importance of social cohesion in these sub-units goes beyond a shared appreciation of the solidarity generated by simply socialising together. In

---

20   Interview 10.
21   Interview 6.
22   Interview 1.
23   Interview 11.
24   Interview 8.
25   Interview 10
26   Bury, 'Recruitment and Retention', 31.

fact, many units described the nature of their social relations in terms of being a 'family'. Interestingly, some of the most cohesive and professional regular infantry units in the British Army also describe themselves as 'family regiments'. This is usually interpreted as indicating dense associative patterns and strong interpersonal bonds across ranks, coupled with long histories of regimental service amongst certain families, and an often regional identity. Regular British family regiments are usually viewed as being highly socially cohesive. Decisively, this can be exclusive of, or complementary to, professional competence. A similar family motif was consistently repeated in interviews with reserve sub-units:

R1: We actually formed a very tight knit family.

R2: A lot of infantry regiments work. They work because –

R3: They are family.

R2: – they are such a good family group ... My old infantry [unit], it was like having 500 brothers.

R1: But it is like being in an infantry regiment here, where you've got that family atmosphere. People looking out for one another ...

R4: I've been here a long time. I've seen a lot of change. But it is one big family. You can share your problems with people.[27]

The above quote is highly illustrative. Firstly, it displays an awareness of the influence dense, familial-like bonds can have in infantry regiments 'that work'; that are effective. Indeed, this NCO is making an explicit association between strong interpersonal ties and effective unit performance. Secondly, these reservists' perception of family appears to be slightly different from the regulars. Their unit's 'tight knit' family is described as being based on 'looking out for one another' and 'sharing problems.' Reservists from other sub-units echoed these sentiments almost exactly:

R2: There is a massive sort of family vibe thing.

R3: It is a proper family job up here.[28]

Basically it's like, I suppose, you can call it somewhat an extended family sort of thing.[29]

I don't see these lots as mates. They're more as family.[30]

---

27 Interview 1.
28 Interview 5.
29 Interview 2.
30 Interview 8.

Interpersonal relationships are therefore frequently described in profoundly social terms; they often surpass civilian friendship to become deeper, familial ties. Crucially, these attributes exactly match those identified in the primary group by Shils and Janowitz. That they were so readily repeated therefore may also indicate the prevalence of this conceptualisation of effective cohesion amongst the sample.

Other reservists remarked on the importance of these relationships to their continued service:

R1: When I look back to mates who I've met through the TA, it is a big thing. I go out and socialise more with people from here than what I do with any of my work colleagues … I love being round the people here.

R2: Aww!

R3: It's not mutual, though. [Laughter].[31]

The first respondent's assertion that his service comrades are his friends that he chooses to socialise with outside of duties is interesting. Indeed, such is the depth of this social/emotional bond that he professes his love for them. This expression results first in tenderness from a female colleague, but is quickly followed with mock ridicule by a male, indicating both the uniqueness of the confession and the unease amongst other group members at directly expressing the depth of their social bonds. This, of course, indicates their importance. Similar observations were made across other sub-units:

R1: Most of my mates now are these guys here. Civvy lads who I went to school with, I say hello to them, but these are my mates.[32]

R1: I quite enjoy coming down here for the friendship and that sort of stuff.

R2: They always have the bar open, even if there's only a few of us. We'll have a chat.

R3: … You've got to get on with who you're working with, haven't you?[33]

While these quotes provide further evidence of the importance that socialising together has in the AR, it is important to note that collective training also plays a major part in generating social solidarity amongst reservists. Training often builds interpersonal bonds through shared hardship and reliance on others; professional training is conducive to social bonding in its own right. Thus, with FR20 pledging to increase training, the bonds generated by socialising within the

---

31 Interview 5.
32 Interview 1.
33 Interview 10.

reserves would be expected to be strengthened further. However, the last quote above is particularly indicative about reserve cohesion. King describes how some Royal Marines' rescued a comrade that they did not like who was pinned down by enemy fire.[34] King argues that this represents the importance of professional comradeship over interpersonal ties; professional infanteers do not need to 'like' each other to perform effectively. Crucially, the reservist above seems to indicate that harmonious social relations are required in his sub-unit to encourage effective performance. Thus, it could be argued that interpersonal bonds are more important in explaining cohesion amongst reserve logisticians than regular infanteers. This, of course, is important for understanding the locus of reserve logistics cohesion and effectiveness. However, there was no evidence to suggest this reliance on social cohesion was correlated with undisciplined or unprofessional behaviour per se.

One simple definition of cohesion is the ability of the group to stay together under stress,[35] and there is ample evidence to suggest that social bonds remain a key motivating factor for remaining in reserve service. The centrality of social cohesion was highlighted by one NCO, who stated: 'I think if there hadn't been a bit of social [life], I would have handed my kit in two years ago.' Thus, the social element of service is often a critically important retention factor. Indeed, it was responsible for this soldier remaining in service despite the organisational frictions he experienced as a result of FR20. Another reservist went further:

> It's the thought that if I do leave, I'm going to be jacking on [letting down] my mates. And you'd miss it because you would want to know what they're doing. And I think that's what keeps you coming back even though it's been really bad.[36]

For this NCO, social bonds are the central reason for his attendance at training, but also for his continued service in the wake of poor experiences of FR20. Crucially, he describes his motivation for remaining in service in terms of a strong sense of social obligation, to avoid 'jacking on my mates'. Moreover, this is explained in emotional rather than professional terms; that he would miss his colleagues. An officer in a different unit elucidated on why reservists under his command frequently attended training: 'They're mates … so if they don't turn up its: "Where were you?"'[37] Thus, complementing the importance of the social aspect of reserve service for reasons for joining, it appears that interpersonal bonds provide a central explanation for these reservists' attendance at training and their longer-term retention. Moreover, the nature of social relations between these reservists has consistently been described as those of friendship, or more profoundly as family.

---

34 King, *The Combat Soldier*, 357–58.
35 Siebold, 'The Science of Military Cohesion', 44.
36 Interview 2.
37 Interview 8.

This locus differs to the cohesion King, Ben-Shalom et al. and Strachan describe in the regular infantry. It is perhaps not too far to suggest that in these reserve logistics units social bonds remain central to cohesion. Indeed, social cohesion appears to be more important in these units than in the regular infantry.

The Decline of the 'Drinking Club'

Referring to civilian groups, Angus Bancroft has outlined how alcohol can be used to heighten 'group cohesion and solidarity',[38] and in a later work on the British officer corps, I outlined the importance of the consumption of alcohol in generating their social solidarity.[39] Walker also noted this amongst TA officers in particular.[40] In fact, the social cohesion-generating function of collective alcohol intake has long been recognised by the British Army, and is reflected in the very cheap alcohol available in messes in most barracks. Given the greater emphasis on social cohesion in the reserves, it is therefore perhaps not surprising that in the past the TA was often perceived as a 'drinking club'. This term was ubiquitous and often used derogatively, but it indicated a view held across the regulars and in some reserve units that socialising in the TA was emphasised over training.

While the 'drinking club' theme has long been a useful social-psychological tool for reservists to distance themselves from the perhaps less professional practices of their past colleagues,[41] it did emerge as a major descriptor amongst AR logisticians for differentiating between pre- and post-FR20 service. The following senior NCO's quote not only highlights the centrality of alcohol to the TA experience in the past, but also supports the arguments made about the importance of individual skill in effective collective performance in logistics units:

> In the days of the National TA, the [reservist] dockers used to come down and unload the ships to give 17 [Port and Maritime regiment – the regulars] some time off. They'd come down for two weeks [their annual training camp], but because they were professional dockers they could unload in half the time 17 could. So each ship would be done by lunchtime and they'd spend the rest of the time in the Corporals' Mess getting hammered. The Mess used to take more in those two weeks than in the other 50.[42]

However, this drinking club ethos was not only confined to logistics units. Other former infantry SNCOs reported similar experiences:

---

38 Bancroft, A. (2009) *Drugs, Intoxication and Society*, Cambridge: Polity, 62.
39 Bury, P. (2017) 'Barossa Night: Cohesion in the British Army Officer Corps', *British Journal of Sociology*, 68(2).
40 Walker, *Reserve Forces*, 105–106.
41 I am indebted to Vince Connelly for this observation. See *Hansard*, House of Commons Debates, 23 March 2005, col. 285WH.
42 Interview 26.

When I first joined in '97, my God, we used to jump on a 4-tonner [truck], go to Thetford and knock back three or four crates of lager, jump off, rock up, harbour, and you're out on exercise all weekend. It was a drinking club.[43]

Both REME and RLC reservists echoed this sentiment:

When I first joined the TA ... it was a bit of a drinking club. We were all out on the lash every weekend, all together, and all having a good time.[44]

There were a lot of lads ... all they were interested in doing was getting in the Sergeants' Mess at night and getting pissed and rocking up the next day stinking of ale. When you'd try to do any trade training they were that rough [they couldn't do it].[45]

Back in the '80s ... it was a drinking club ... When I was with the regulars, the TA was scum.[46]

Clearly, a widespread perception exists that TA activity was often centred on alcohol to the detriment of training. Socialising with alcohol was perceived as being more of a priority in the past. This, of course, provided easy means by which the professional regulars – themselves no strangers to drinking – could denigrate their TA rivals. Indeed, the prevalence of this motif for the pre-FR20 reserves indicates how the 'drinking club' became a label for describing the organisation's perceived unprofessionalism. Crucially, however, its widespread use today allows current reservists to distance themselves from this perceived unprofessionalism, while also highlighting how reservists are internalising regular values. This indicates the latter's importance in setting norms in today's Army Reserve.

Despite the continued existence of the drinking club metaphor, there is much evidence that its validity is waning. There are many contributing factors to this, including the closure of AR barracks and the enforcement of drink driving legislation, but amongst the sample, the primary reason was perceived to be a result of gradual TA professionalisation since 2003 which has been intensified by FR20:

Mod: Do you think it's [the drinking club ethos] changing in the reserves?

All: Yes.

R1: Big style.[47]

---

43 Interview 1.
44 Interview 5.
45 Interview 1.
46 Interview 2.
47 Interview 8.

R2: It was a drinking club. Now it's not.[48]

For these respondents, FR20 has been critical to the decline of the drinking club ethos. Other experienced SNCOs directly made the link between it and increasing professionalism:

The level that you have got to be at, the [professional] standard, it's not a drinking club now, which it used to be.[49]

It has moved on somewhat in the last 30 years and it is a more professional unit than it used to be. But it is still down to the attitude of the people of the regular army.[50]

It wasn't just the TA [that was a drinking club]. The regulars were just the same. For one reason alone: that was the culture that we lived in.

The last quote is especially informative as it attempts to explain the drinking ethos as a reflection of the dominant culture at that time in both the regular army and society. Despite the prevalence of the drinking club motif, numerous interviews indicated that while this was seen as a valid label to describe the TA's lack of professionalism in general, it was not applicable to their sub-units in particular.[51] This not only highlights the sensitivity to being tarnished; it also indicates their desire to be viewed as professionals.

While FR20's increased training standards and the wider policy drive to reduce drinking in the regulars have contributed to the decline of the drinking club,[52] there was a recognition that this culture needed to be explicitly addressed for this to happen. As a result of FR20, some unit Commanding Officers have issued new rules designating certain training weekends completely 'dry'. As one officer commented: 'We have definitely professionalised as a unit. With the drinking club, there was a culture to break, and we have … The message went down that you'd better be ready for duty in the morning or standby.'[53] A senior officer elucidated that he had introduced a 'dry weekend rule' in order to underline that reserve service in his unit would not be tied to the consumption of alcohol.[54] Another officer commented on the effect of this: 'It is changing. And if that means some of the old and bold turn to the right and march off, then so be it … maybe that's not such a bad thing.'[55]

---

48 Interview 8.
49 Interview 7.
50 Interview 2.
51 Interviews 5, 7, 24.
52 *Soldier Magazine*, August 2016, 12.
53 Interview 23.
54 Interview 26.
55 Interview 27.

Other sub-units reported a decline in social events since FR20. Whether this was intentional or not was not revealed, but most reservists accepted it as the price of professionalisation; there is now less time for social events as the focus is on training.[56]

Although they are widely viewed as complementary and – in the TA's case – intertwined, there is obviously a distinction between social cohesion and alcohol consumption. This was made by a number of reservists. One NCO stated that the Army Reserve of today 'is a social club mainly, not a drinking club'.[57] Another infantry officer elucidated:

> I haven't noticed a decrease per se, there's still the social side to army drinking, but a lot of time after training we'd go to the bar and everyone has a Coke ... the bar's ambience and the extra time we have there on top of training is good for getting J1 [personnel issues] done. So in terms of social cohesion, it's important, even if people aren't drinking.[58]

The Mess therefore remains an important site for generating and sustaining social solidarity, even though the consumption of alcohol is less frequently the focus. Indeed, without alcohol, the distinction between Mess activity and professional duty appears to be increasingly blurred, allowing as it does administrative issues to be addressed to enhance sub-unit effectiveness. Supporting this, another officer spoke of how 'a lot of networking is done up in the messes'.[59] As such, the 'drinking club' may be in decline, but the club itself remains profoundly social, and indeed, professionally useful, as – it must be noted – does the motif.

The Rise of Professionalism

Asked whether social cohesion had previously been based around the consumption of alcohol pre-FR20, one NCO stated: 'I think it was ... the new way – the [regular] army way – is about courses.'[60] While this quote underscores the last chapter's findings on the increased availability of courses as a result of FR20, it also points to something more profound. This chapter has already shown how FR20's drive for professionalisation has resulted in a concerted move away from alcohol-aided social cohesion. As the NCO above suggests, these reservists' new professionalism is increasingly based on that of the regulars; on training and competency. The processes by which reserve logistics sub-units are professionalising, and their

---

56   Interviews 5, 10.
57   Interview 24.
58   Interview 24.
59   Interview 11.
60   Interview 25.

implications, are closely related to understanding cohesion in these sub-units, and the overall impact of FR20. They are therefore worthy of further examination here.

In these reserve units, qualifications gained through attending courses, and importantly, operational experience, are increasingly viewed as the standard by which individual reservists judge themselves *vis a vis* the regulars. This new, more professional attitude appears to have gradually percolated into the TA during the increased deployment of reservists on operations in the 2000s. It has also been boosted by the marked increase in opportunities for all reservists to train and deploy with regulars.[61] This sentiment was echoed in other sub-units and across ranks, as the quote below shows: 'When I was in the regulars years ago – when it was the Cold War – the TA was a joke. It's got more professional since 2003.'[62] Again, this reservist's disassociation from the old, unprofessional TA is notable, but, underpinning FR20, the impact of post 9/11 operations on reserve professionalism is clearly viewed as a critical source of transformation. As such, at an organisational level, deploying with the regulars has professionalised the reserves in and of itself.

However, perhaps most importantly, the increased exposure to the regulars on operations has imbued a growing perception amongst these reservists of their service being a job. The government's 1978 Shapland Report stated that although TA service in the organisation was a 'demanding hobby', it was a hobby nonetheless.[63] In the late 1980s senior officers also admitted to Walker that reserve service was a distant third priority after family and work.[64] While that order of priority may remain, Connelly also outlined numerous professional and cultural barriers to integration between the regulars and the reserves, including reservists' perceived lack of professionalism. However, I contend that the conception of service is gradually changing. This is occurring in two distinct ways. Firstly, with the increased training burden, reserve service is now viewed as a job, albeit usually a part-time one, with the accompanying level of commitment and attention to detail required:

> R1: Even in the six years that I've been in, it's gone from being a hobby to being a part-time job. You wouldn't miss your full-time job so you can't miss your part-time job either.

> R2: ... [It's part-time but] It's still a job.[65]

In tandem with the decline of the drinking club, reserve service – despite being part-time and more reliant on social cohesion than the regulars – is not viewed

---

61 Interview 7.
62 Interview 5.
63 Walker, *Reserve Forces*, 6.
64 Ibid., 44.
65 Interview 5.

as a hobby anymore. The second distinct way the reserves are drawing closer to the regulars concerns individual reservists' performance. Increasingly, these reservists must be individually competent enough to 'do their job'. The importance of individual competency in order to work alongside the regulars safely and effectively was repeatedly emphasised:

> You have to be at that standard. You cannot think that you can rock up to a [regular] unit [not at their standard], because you've got to remember who you are representing as well.[66]

> You need to prove to yourself you can do the job.[67]

Thus, the ability to fulfil their role competently on operations alongside the regulars is the ultimate standard by which these reservists judged their own professionalism. Another reservist explicitly commented on how the regulars' performance acted as the yardstick by which professional competency is measured: 'You test yourself a bit ... in your mind they [reservists] benchmark against someone of the regular rank equivalent.'[68] While the fact that reservists could never meet regular capacity due to their part-time nature was recognised,[69] doing your individual job to the standard of the regulars is now seen as the benchmark for competency in training and on operations. For most of its members, the Army Reserve is now a part-time job. For others who have more time to take advantage of the increased opportunities to train and deploy – especially those without civilian employment – it is their full-time job, raising interesting questions about a growing core-periphery divide discussed in the next chapter.

In doing your job, personal and collective status is at stake. Supporting King's observations on status honour in the infantry, there are signs that these reservists are becoming increasingly aware of the need to earn and maintain their professional status by the measures defined by the regulars. While sensitivity about the regulars' perception of reserve professionalism existed before the 2003 deployments, it appears that greater exposure to the regulars on operations and as a result of FR20 has increased this sensitivity. Across the sample there was a desire to be viewed as being as capable as the regulars. Again this was usually compared with reference to individual performance on operations. This desire to match the regulars' professionalism to maintain their own status and that of their sub-unit, and the reputation of the Army Reserve, was clearly put by an ex-regular: 'We have to be at that standard because otherwise we are letting ourselves down.'[70] Indeed, professional reputation appears to be very important:

---

66 Interview 7.
67 Interview 6.
68 Interview 1.
69 Interview 2.
70 Interview 7.

R1: They [the regulars] hate us. We've got such a bad name because of these guys.

R2: Because of the old [sub-unit] was so dodgy and cut corners ... they made so many mistakes.

R3: They coated [Camp] Bastion in fuel.

R1: Now we're [new sub-unit], people are like, 'Oh, who are you with?' 'We're with [new sub-unit]', all excited and happy [but] they're like: 'Err ... fuck off', because of stuff that happened. We've inherited their bad reputation.[71]

This quote not only highlights a perceived lack of professionalism before FR20, but also an increasing sensitivity to the enduring impact of bad reputation as closer integration is undertaken in its wake. Another interview revealed an incident where a regular infantry platoon commander had been forced to move two reservists out of his unit for their consistent failure to perform to the expected standard on patrols in Afghanistan. These soldiers' conduct embarrassed their fellow reservist colleagues.[72] As these experiences show, reputation matters deeply to reservists.

The surveys also provided complementary data in terms of professionalisation. Forty-four per cent of respondents agreed that they 'worried a lot about meeting the expected professional standard in their sub-units', with 33 per cent disagreeing. By comparison, a regular infantry sub-unit recorded 50 per cent in the category, and 22 per cent respectively. Previous research has identified that internalised, highly professional values are often a major source of personal anxiety in elite regular units. Thus, this finding is interesting as an expected – albeit minor – difference between reserves and regulars is observable. Meanwhile, 52 per cent of reservists agreed that 'it was more important to be a good soldier than to be liked', with only 15 per cent in the disagree category, indicating the importance of professional values. This figure almost matched that of a regular infantry sub-unit (54 per cent agreement, 10 per cent disagreement). It was supported by another question concerning the risk of 'deviant cohesion' that Donna Winslow identified can be a problem in units with high levels of social cohesion and not enough discipline.[73] Asked whether it was more important 'to be 'one of the lads' than a good soldier' in their sub-unit, 68 per cent disagreed, and only 12 per cent agreed. There therefore appears to be a strong awareness that social cohesion must be balanced by professional values. Thus, it appears that these reservists are experiencing the gradual permeation of the regulars' ethos and norms into their domain. The transformation is gradually

---

71 Interview 1.
72 Interview 5.
73 Winslow, D. (1999) 'Rites of Passage and Group Bonding in the Canadian Airborne', *Armed Forces and Society*, 25(3).

changing these reservists' attitudes, bringing them in line with those of the full-time professionals. Nevertheless, for the majority of these reservists, the regulars are the final arbiters of their professionalism.

Given this widespread sensitivity to regular perceptions, the potential damage to individual and collective status caused by poor performance can be expected to motivate reservists in a similar manner to the status honour King observed. While this may still be much more pronounced in the infantry, it does appear that most reservists have become more sensitive to their professional reputation as a result of closer integration with the regulars. Evidence of this was also revealed in reservists' positive experiences on operations, where being mistaken for a regular was a recurring story told to indicate individual competence. One officer recounted a regular colleague asking if a reservist under the former's command was TA. When he replied that she was, he received the comment: 'I just automatically thought she was so professional that she was a regular.'[74] Accounts such as this were repeated frequently, and were always told with palpable pride. For example: 'The guys that I was on tour with didn't even realise I was TA until about a month before we were leaving theatre, and they went, 'What, you are TA? I didn't realise you was TA.'[75] Professionally therefore, the best thing for a reservist is to be mistaken for a regular on operations due to their performance. Meeting or exceeding the performance expected of a regular on operations is thus the gold standard for a reservist. Thus, in performing like a regular, a reservist can become an honorary professional. Moreover, this hard-earned professional status is not always temporary. Reputation has been established and as long as it is maintained by performance, it will continue to be acknowledged in the reservist's network. This was revealed by many reservists who cited the much more welcoming response from regulars with whom they had deployed and hence proved themselves to. For these individuals, their hard-earned reputations allow them access to the coveted professional status group policed by the regulars. Similarly, as we have seen, poor performance can result in long-term ostracism by the regulars.

By better integrating the reserves with the regulars, FR20 is slowly instilling a professional culture. This is evidenced by these reservists' desire to distance themselves from the drinking club motif; an increasing awareness of the standards expected on operations; the view that reserve service is no longer a hobby; and greater sensitivity about their performance-based personal and collective professional reputation which must be earned and maintained by competent performance. In short, these reservists increasingly see the regulars as the benchmark for performance and seek their approval to confirm their status. However, while these processes indicate the reserves' rising professionalism, there is ample evidence to suggest that it has not met the standard the regulars

---

74 Interview 3.
75 Interviews 1, 7.

expect. Supporting Connelly's recent findings on the limits of integration, reservists found regular attitudes to them as frequently less than positive, but slowly changing:

R1: I deployed with the infantry in 2007. Trust me. It hasn't changed.

R2: It was a lot worse, I think.[76]

They hate us.[77]

R2: I think it's going to take a lot longer for the regular army to recognise the reserve. To see how competent they are. It's going to take a few more years.

R3: ... [But it] has changed a lot over the last couple of years.[78]

Some were much more positive, stating: 'They were brilliant to us' or 'It does work. There's a mutual respect there',[79] while numerous others pointed out that the success of sub-unit integration depended heavily on the command culture in the units concerned and on personal relationships.[80] As with sub-units' experiences of FR20, the mixed responses support the importance of the command environment in this regard. As such, despite progress, there appears to be some way to go before reservists perceive regular attitudes to them to have changed considerably. Indeed, this is supported by the results of the 2018 ResCAS survey which found that only 31 per cent of Army Reservists felt valued by the regulars.[81]

Nevertheless, despite gradual professionalisation, the importance of social cohesion still persists. In order to examine the extent to which sub-units had professionalised, respondents were asked the blunt but pertinent question if 'unit members viewed each other as professionals or mates first?' The responses below are indicative:

All: Mates.

R1: I think that's one of the biggest differences between the [reserves and regulars].[82]

R1: Mates.

R2: Family.

R3: Family, yes.

---

76 Interview 5.
77 Interview 1.
78 Interview 6.
79 Interviews 3, 7.
80 Interviews 2, 3, 7.
81 Ministry of Defence (2018) *Reserve Continuous Attitudes Survey June 2018*, 8.
82 Ex-regulars, in interviews 5, 8.

R4: Family, definitely.

R5: Yes, definitely.[83]

Mod: Would you be mates with someone who is below the expected standard of their rank and experience?

R1: No.

R2: Yeah, we are. The thing is, we all have to help each other out.

R3: It's still a family.

R4: ... You work as a family. I think. Once you're in here, you're family.[84]

R1: No. We'd be friends.

R2: But not in the same way.[85]

Other reservists were asked whether they would be friends with a colleague who was 'not up to the required professional standard'; 70 per cent agreed that they would, highlighting the social locus of cohesion in these units.[86] While the same percentage was recorded in a regular logistics sub-unit, in a regular infantry sub-unit this was only 50 per cent. The overwhelming majority of reserve logisticians, and an indicative sample of regular logisticians, continue to describe the nature of the relationships with their comrades in profoundly more social terms than the infantry. This suggests that despite ongoing professionalisation, social cohesion remains central to understanding the associative patterns between most of these reservists. But as the quote above indicates, there were some who did not view their relations in this manner, but in the more professional terms that have been identified in the regular infantry, where failure can result in immediate ostracism.[87] Indeed, amongst higher ranks with more responsibility there were more qualified, if broadly similar, responses:

R1: That's a fine line because some of us have worked together 20 years.

R2: In this room, mates.

R3: Can you not switch off from being a mate to a professional soldier when you need to?

R3: I can be friends with so-and-so, and when it comes to Friday night we turn into a professional soldier.[88]

83 Interview 7.
84 Interview 10.
85 Interview 1.
86 Survey data collected 2015.
87 King, *The Combat Soldier*, 347.
88 Interview 6.

It depends what the situation is. If we're together in bar, it's mates. But when we're here on a Wednesday night, and we're doing stuff, it is professionals. It's got to be professionals. Ex-regular and all that, you've got to be professional.[89]

This understanding of professionalism as situationally-dependent is not unexpected given the literature on how reservists negotiate their identity. But it does indicate that notions of professionalism amongst these soldiers, and hence the sources of cohesion, are perhaps more fluid than in regular forces. It appears that even amongst these senior reservists and ex-regulars, interpersonal bonds still remain important, and that professionalism can be switched on and off depending on context.

Discipline

In *The Combat Soldier*, King explicitly links the rise of professionalism in regular armies with a change in the source of their discipline. King argues that Western conscript armies of the early-mid twentieth century relied heavily on the threat of punishment to maintain battlefield discipline and encourage combat performance.[90] In modern professional armies, however, he argues that organisational effectiveness is much more reliant on soldiers' self-discipline. This is due to the increasing importance of status honour, and the threat of professional shame amongst volunteer soldiers who do not perform to the expected standard. In contrast to a lack of official punishment, King details the often serious sanctions applied to group members who fail to perform effectively, like a Parachute Regiment soldier ostracised for poor performance in training and a Royal Marine publicly ridiculed for similar conduct in combat.[91] For King, the intensive collective training conducive to heightened professional obligations between soldiers has changed the nature of military discipline in some Western armies. Indeed, in a later publication he contended that 'A new paradigm of military discipline seems to have emerged'.[92] This new paradigm is reliant on core groups of comrades united by professionalism and enduring social relations to motivate correct performance. According to King, the primary power of professional self-discipline amongst combat soldiers is strengthened by dense social bonds in these core groups. Nonetheless, it is important to stress here that in these combat core groups, friendship is subordinated to professionalism.

---

89 Interview 11.
90 King, *The Combat Soldier*, 362–75.
91 King, A. (2015) 'Discipline and Punish: Encouraging Combat Performance in the Citizen and Professional Army', in King, A. (ed.) *Frontline: Combat and Cohesion in the Twenty-First Century*, Oxford: Oxford University Press.
92 Ibid., 112.

Another important work here is Thomas Thornborrow and Andrew Brown's fascinating case study of identity and discipline in the British Parachute Regiment. Like King, for these authors the aspiration to be an elite 'Para', and the threat of professional and social ostracism for failing to meet the expected standards, are central to explaining the high levels of self-discipline and individual performance in the unit.[93] Crucially, they show how the desire to conform to the heightened professional behaviour expected in elite regular units functions like a Foucauldian panopticon to monitor interactions and encourage performance.[94] Echoing King's observations about the threat of group ostracism for performance failures, this occurs to such an extent amongst the Paras that even experienced senior ranks reported status anxieties during routine duties.[95] Thus, for both King, and Thornborrow and Brown, self-discipline and surveillance – both based on professional competence – are crucial to understanding performance in infantry units.

However, neither of these important works addresses the nature of discipline in non-infantry forces, nor in a reserve force that is gradually professionalising. In 1990, Walker noted that, due its volunteer ethos, TA discipline was 'lax' and 'ad hoc' compared to the regulars.[96] One TA officer's comment at this time is illustrative: 'there was a dull indifference to discipline, but a wonderful loyalty to duty'.[97] This quote indicates that at this time formal punishments were rarely resorted to and instead there was a high degree of self-discipline. However, the decisive fact that – despite ongoing integration with the regulars, and in contrast to them – every Army Reserve parade is still voluntary suggests that the nature of discipline may still be somewhat different from the regulars. The lower enforcement of military discipline in the reserves was frequently commented upon by reservists, and nearly always in reference to the application of Army General Administrative Instruction (AGAI) 67 system which governs both regular and reserve forces. For example:

The AGAI system is there, but we don't need to use it.[98]

R1: I think that's [discipline] one of the biggest differences between the two.

R2: It's much more relaxed, yes.

R3: More relaxed but the job still does get done.[99]

---

93 Thornborrow, T. and Brown, A. (2009) 'Being Regimented: Aspiration and Identity Work in the British Parachute Regiment', *Organization Studies*, 30(4): 364–65, 367–68.
94 Thornborrow and Brown, 'Being Regimented'.
95 Ibid., 365.
96 Walker, *Reserve Forces*, 71.
97 Ibid.
98 Interview 2.
99 Interview 8.

> If you try and do discipline like you would in the [regular] battalion, you
> would not have people turn up.[100]

These groups consisted of some ex-regulars, adding credibility to their claims that
reserve discipline is different. Moreover, AGAI 67 action was widely perceived
by these reservists to be resorted to much more frequently in the regulars as
their contractual and legal obligations compelled them to duty and discipline
in a way that, while also applicable to reservists, is unenforceable in reality. A
reserve unit's regular adjutant (the officer responsible for discipline) supported
these claims:

> Discipline is different … the AGAI system is a blunt sword to be honest. The
> thing is, every parade is a voluntary one, and using the discipline system is
> contrary to what you are trying to achieve. But it is used and we do use it.
> You don't get it for in-subordination or [bad] turnout; those incidents occur
> less than in the regs. To be honest, most of the discipline issues are alcohol-
> related, like the regulars.[101]

Other reservists noted the difference between reserve infantry and logistics units.

> R1: Discipline's slightly different. The infantry was more disciplined. Here
> it's …
>
> R2: A bit more laid-back.
>
> R3: Relaxed.
>
> R1: To be honest with you, sometimes you'd rather have the infantry here …
> you knew where you stood.[102]

Contrasting King and Thornborrow and Brown, these views indicate that logistics
reservists still perceive discipline in the regulars, and in the infantry, in the
traditional terms of punishment, rather than the emergent paradigm of professional
self-discipline.

If the punishment system is not as frequently resorted to in the reserves, then how
is discipline maintained? Unsurprisingly, the lack of use of the official discipline
system compared to the regulars was consistently normalised by reference to the
reserves' own distinctive discipline. For instance:

> It's a different mentality. I've got recruits, I can't drag them around the floor
> because they won't come in. I mean, they do things wrong. I'm not one for

---

100 Interview 10.
101 Interview 23.
102 Interview 5.

shouting and bawling at people. I don't like being shouted and bawled at at work.[103]

R1: That's always been the best thing about the TA, isn't it?

R2: ... That's the difference in mentality in the TA and the regs. A regs bloke would be like: 'Get a fucking grip, sort your life out,' and beast them until they get it right.

R3: ... There's much more competition for promotion in the regs. Guys want to look better than somebody else.[104]

For these reservists, discipline is perceived as reflecting a different mentality based on the identity of being a volunteer. It is also enshrined in the different regular/reserve terms of service, whereby reservists are not punished for missing parades. Both are conducive to a reluctance, perhaps more prevalent in logistics units, to robustly enforce standards through verbal and physical punishments traditionally associated with regular discipline. As the last quote suggests, the lack of professional competition in the reserves is one reason for this. Another is that, without fully enforceable punishment, the social bonds of friendship are much more important in encouraging correct behaviour and, ultimately, turnout in the reserves. This highlights the recognition that resorting to the discipline system indicates the breakdown of social harmony which is crucial for working relationships. The quote below is instructive:

R1: If one of your brothers makes a FUBAR [Fuck Up Beyond All Recognition], how do you deal with it?

Mod: You're going to tell him.

R1: And that's what it's like here. Sort of: 'Come on. Don't let the rest of the family down', sort of thing ... You don't want the rest of the family to suffer for a simple mistake ...[105]

Maintaining the reputation of the unit is notable here, as is the deeply social terms in which this reputation is conceived. But related to this, the respondent is also suggesting that the informal enactment of social bonds – of social obligation – is used to encourage performance, The reservist core group, based on the social bonds between 'brothers' or 'family', appears very important in doing so. Indeed, another NCO stated: 'We tend to try and sort things out at a low enough level';[106]

---

103 Interview 10.
104 Interview 8.
105 Interview 2.
106 Ibid.

that colleagues discipline each other informally for bad behaviour. These points are informative as they suggest that reserve discipline may in fact be closer to that of the professional regulars than the reservists above initially acknowledged. In a slightly different way to that observed by King and Thornborrow in elite infantry, it appears that primarily the power of interpersonal – rather than professional – bonds encourages performance in the British reserves. The vector of self-discipline is different, but the process the same.

However, reservists were cognisant of the fact that the need for social solidarity has its dangers. 'I think we should AGAI more people because the discipline sometimes is a bit close … Because we are volunteers and we all know each other, sometimes we do get a bit wishy-washy.' Similarly, they noted how, like the regulars, informal punishments were much less robust, or 'colourful', than in the Cold War-era TA.[107] Nevertheless, there was a general sense that, despite the reluctance to use the discipline system, the reserves are still relatively disciplined. According to one commander: 'There are far less discipline problems … [but conversely] there's less discipline in the reserves, and that's not because we're shying away from it, it's generally because the guys want to be here, they volunteer to come here, and they don't want to burn their bridges.'[108] This quote again highlights the perception that reservists who volunteer for each parade generally have better self-discipline compared to some regulars. While this is in fact questionable as more lax conditions of service could be perceived as conducive to lax discipline,[109] given the importance of social cohesion in these units, it was interesting to note how ex-regulars related this trust-based discipline to social events involving alcohol:

> I think there's a little bit of a trust element as well. Generally the age of the reservist is slightly older than a regular soldier. They know their boundaries, particularly with drink …[110]

As another regular officer commented: 'Yes, we can trust them … they don't take the piss as much.'[111] Trust, therefore, seems to play an important part in explaining discipline in the reserves. While this trust may rest more on social rather than professional obligations – and it is certainly different from Ben-Shalom et al.'s 'swift trust' generated in diverse regular units by standardised training procedures – it nonetheless exists. Indeed, reservists' recognised maturity also indicates the central importance of self-discipline in building this trust.

The importance of this maturity in explaining reserve discipline was also evident in how these reservists dealt with unit members who performed badly. As

---

107 Interview 2.
108 Interview 3.
109 I am indebted to Alex Neads for this insight.
110 Interview 6.
111 Interview 3.

detailed by Thornborrow and Brown, and King, in the regulars the punishment for poor performance is frequently social and professional ostracism from the core group. Professional status brings with it the threat of professional shame. In the less professionalised reserves, the opposite appears to be the case. Faced with less time to train and therefore varying degrees of skill, instead of excluding failing members, most of these reservists viewed such individuals as the target for development and encouragement rather than ire and exclusion:

> R1: We always help each other out. If someone's struggling with something
> …
>
> R2: Find out why.
>
> R3: Help them.[112]
>
> You help them out, don't you?[113]

Clearly then, unlike their regular colleagues, there is an innate awareness amongst reservists of their professional limitations. As the costs of professional failure are not as high as in the regulars, a more conciliatory approach focused on rehabilitation and mentoring is usually followed. This is in stark contrast to some regular units.

Taken together, the lack of use of the AGAI system, the different mentality of self-discipline and trust, and the more conciliatory approach to poor performance suggests that discipline in these units is different to that identified in the regular infantry. This is perhaps unsurprising given the former's reliance on social cohesion; as a result, interpersonal bonds likely have greater regulatory power than in the regulars. But this situation also stands in contrast to discipline in conscript armies, which although they relied heavily on social cohesion, used the punishment system frequently. Drawing on the evidence presented above, perhaps it is not too much to theorise that reserve discipline appears to be a hybrid of that discussed in the cohesion literature, blending the new professional paradigm of self-discipline with a reliance on social cohesion's interpersonal relationships for enforcement. Hence the paradox in reserve discipline: on the one hand, the AGAI system is not used as it destroys the social cohesion upon which discipline relies. On the other, it is not needed precisely because social bonds act as a disciplining mechanism when self-discipline has failed. While FR20 is likely to gradually increase the strength of professional shame through better integration with the regulars, given this distinct nature of reserve discipline, it is unlikely to fully solve this paradox in the short term. Nor, as the evidence suggests, does it need to.

---

112 Interview 8.
113 Interview 7.

Conclusion

This chapter first examined perceptions of cohesion and readiness, and then the nature of sub-unit culture in terms of cohesion, professionalism, and discipline in order to determine the impact of FR20. It found that perceptions of cohesion and readiness remained relatively high in 2015–2016, but that FR20 was not significantly increasing these either. It also found confidence in FR20 was decreasing over time. Moreover, in logistics units that require collective performance, cohesion remains similar to that in the professional infantry and is based on personal and collective drills. Successful execution engenders successful group performance. In singleton logistics trades, cohesion is viewed in predominantly social terms, distinct from performance. Despite this important distinction between trades, overall it is clear that social cohesion remains central to explanations of why soldiers join, attend training, and remain in the reserves. It also encourages performance. Given that reservists have less time united as a group, social cohesion therefore appears to be more important in the reserves than in the regulars, and it is likely – and indeed wise – for it to remain so. However, its nature is changing as a result of FR20. In the past it was based on alcohol, but the drive for professionalism has led to the decline of the drinking club. More frequently, collective socialising does not always involve drinking. Conversely, professional cohesion, based on training and courses, is growing. Due to greater exposure to the regulars in training and on operations, the reserves are gradually professionalising by emulating their regular colleagues' culture and attitudes towards competence. This is most obvious in reservists' conceptions of their service as a job, their recognition that operational effectiveness represents the standard at which they should be judged, and their widespread acceptance that the regulars have the right to judge them. Their growing sensitivity to individual and collective reputational damage, and their desire for professional status, provide further indicators of the slow percolation of professional culture into the reserves as integration with the regulars continues under FR20. As discussed in the next chapter, it is important for the reserves' long-term survival that this occurs by emulation and integration rather than assimilation. Meanwhile, reserve discipline occupies a unique position between the professional and social worlds, reliant as it is on both self-discipline and social bonds. Given the voluntary nature of reserve service, this is unlikely to change anytime soon. The same is true of the social source of cohesion in these sub-units. Nevertheless, FR20 is gradually encouraging a more professional culture in these reserve units. Given the serious issues concerning their ability to deliver the required capability, it may be that this becomes one of FR20's most enduring successes.

Finally, in terms of the transformation literature, this chapter has shown how the Army Reserve's cultural emulation of the regulars has occurred at, and been shaped by, micro-organisational factors. In trying to emulate the regulars' competency-based professionalism, reservists have begun to gradually change their traditional organisational culture and adopt some of the attributes of the regulars that define them at the micro-level. However, the part-time nature of reserve service, the resulting differences in the locus of cohesion between reservists and regulars, and the nature of singleton logistics trades has limited their ability to fully emulate the regulars. The differences in the associative patterns that generate and sustain cohesion and professionalism between the regulars and the reserves has therefore shown the importance of bottom-up factors in influencing cultural transformation. It seems obvious, but the micro-level organisational reality of different cohesive sources, themselves related to organisational nature, can limit the degree to which military organisations are able to emulate those they wish to.

Chapter 7

FR20, Transformation and Society

This book set out to examine the historical, organisational and conceptual origins of FR20; these origins' impact on FR20 on reserve logistics sub-units; and what this outcome tells us about professionalism and cohesion in reserve units and the wider transformation of the Army Reserve. Complementing the post-Fordist conceptual approach, a number of other perspectives and methods were used to examine different aspects relating to FR20. Here, a brief summary is perhaps worthwhile. In the second chapter, I examined the past periods of top-down reserve reform to show how these were cyclically influenced by similar economic, strategic and recruitment factors. These were usually accompanied by politico-ideological arguments to support the organisational changes that followed, while the actual reforms themselves were often hindered and delayed due to stakeholder resistance and organisational frictions in both the army and the reserves. Following Allison and Kaufmann, I therefore argued that transforming the reserves has proven a difficult endeavour historically.

Chapter 3 provided the post-Fordist organisational context for explaining FR20's impact on reserve logistics sub-units. In doing so, I showed how post-Fordist principles that have informed, and the practices and processes which have shaped, the transformation of US and British military logistics. I argued this post-Fordist transformation has occurred primarily at the strategic and operational management and structural levels, but that tactical logistics practices have also adapted. In doing so, I challenged the classical literature on military logistics for being in many respects out of date. Decisively, I showed how the post-Fordist approach has been absorbed beyond simply military logistics functions to shape wider British and Western force structures, including Army2020 and FR20, and, in their drive for efficiencies, had also increased their potential vulnerability to strategic shocks. In Chapter 4, I examined the origins and evolution of FR20 policy, detailing how it emerged in the backbenches of the Conservative Party and was supported by elites for their own intra-party political reasons. These intra-party political origins resulted in heavy resistance from the army's leadership to the transformation and an opportunistic plan that, despite numerous revisions during the policy formulation process, resulted in an overly ambitious, unplanned, top-down attempt at transformation. However, by including the reserves in the army's deployment schedule, FR20's overarching political *raison d'être* was met; the survival of the TA was all but guaranteed.

Ultimately, as Chapter 5 discussed, the outsourcing of logistics capabilities usually held in the regulars as a result of Army2020 and FR20 increased the burden on, and created major organisational frictions in, many of the logistics sub-units examined. Some of these had to undergo profound transformation as a result of the new policy, including changing roles and locations, or being formed from scratch. Positive impacts of FR20 have been revealed, including the increased availability of professional courses, and greater opportunities to deploy and train with the regulars. But there is also evidence of frustration with the overemphasis on recruiting activity, concerns about recruit quality, and of units being 'forced to play politics' over their recruitment figures. Meanwhile, the post-Fordist centralisation of equipment stores and defence estate rationalisation were found to be negatively impacting reserve training activities. Most importantly, however, combined with poor recruitment figures, given many of the sub-units' need to re-role and the very heavy re-training burden this created, evidence was provided that many of these sub-units did not believe they could meet the capability required of them under Army2020. Thus, extensive organisational frictions were undermining FR20's drive to outsource hard military capability to these units.

Chapter 6 examined Army Reserve cohesion from both the quantitative Standard Model approach and that of the qualitative revisionists. It found that perceptions of cohesion and readiness amongst logisticians were positive but had not changed significantly as FR20 progressed. Importantly, over time there were significantly lower levels of confidence in FR20 increasing their sub-units' capability. Conversely, the policy had made progress in increasing reservists' exposure to the regulars. These findings supported the argument that FR20 is failing to deliver hard capability, but is delivering better integration with the regulars in the sub-units examined. More broadly, the evidence also indicated that perceptions of social cohesion amongst reserve logisticians were higher than in regular units, who, conversely, had higher task cohesion scores. This supported my later argument that the nature of reserve logistics cohesion is different to that of the regular infantry.

The chapter also examined the nature of logistics and reserve cohesion using the selected sub-units as an evidential base in order to assess FR20's impact on the cultural-normative aspects of cohesion, professionalism and discipline. At the tactical level, apart from an increased infantry training requirement for logisticians operating on unsecure lines of communication, post-Fordist logistics organisation was found to have had relatively minor impact on skills, but there was an acknowledgement that future technologies could change current structures, specialisms and individual skills considerably. Similarly, the evidence suggests that social bonds remain a crucial reason for reservists joining, attending training, and remaining in service, and that reservists classify the nature of their relationships with their comrades in the terms of 'mates' and 'family'. Despite the continued importance of social cohesion, as a result of increased deployments since 2003 and FR20, it is clear that there has been a significant decline in the drinking club and an important rise in professional culture and ethos. This professionalism was acknowledged to emulate, and is closely policed by, the regulars, and was evidenced in the increasing understanding of reserve service as a part-time job where both individual and collective professional status was at stake. As such, FR20 is having a major impact on the professional culture of these units, but perhaps less so on the fundamental nature of their cohesion due to the differing nature of reserve service. Finally, I used this evidence to posit that inherent bottom-up, micro-level factors such as the source of cohesion can limit the ability of military organisations to emulate those they wish to. This has broader implications for the transformation literature.

However, there are broader conclusions to be drawn from these arguments which are worthy of exploration here. Firstly, I examine some of the wider organisational impacts of FR20, and what this may mean for reserve service in the future. I also discuss how the army has eventually reconciled a externally-imposed and problematic transformation with the political desire to reinvigorate the Army Reserve. Secondly, I discuss what FR20's 'partial transformation' of logistics sub-units tells us about the transformation literature. I then analyse the experience of FR20 in terms of recent British civil–military relations, before finally drawing some wider conclusions about FR20 and modern British society.

An Emerging Division

The enduring nature of reserve service has been described as an 'equilateral triangle' characterised by commitment to the military, family, and employment by both Carter and Wall.[1] Both are therefore keenly aware of the need to carefully balance these in order to maximise the reserve's enduring contribution to Britain's overall military capability. Indicating the importance of this triangle, other research projects have examined how reservists balance work and family life and manage identities as a result of FR20. Indeed, Edmunds et al. have already identified the increasing demands of the post-FR20 reserves on its member's time, labelling it one side of an 'iron triangle of greedy institutions.'[2] As a result, these authors call for greater support from the military for families and employers in the wake of a more operational role for the Army Reserve. But crucially, they do not discuss the wider impact this integration and operationalisation of the reserves is having within the Army Reserve itself.

In fact, there is evidence to suggest that the greater individual and collective training burden placed on the reserves is altering the nature of Army Reserve service itself. This change is being driven by a number of factors related to FR20: the growth

---

1   Interview, Wall, 10 May 2016; Interview, Carter, 11 May 2016.
2   Edmunds et al., 'Reserve forces', 131.

of Full-time Reserve Service posts; the opening of regular posts to reservists; the better availability of courses and deployments; and the increased sensitivity to professional status due to integration with the regulars. Complementing it, and very important in areas of low employment, has been the availability of better monetary rewards for reservists who are not in full-time employment. This greater demand on reservists' time is threatening the equilateral triangle to such a degree that it is challenging the traditional, part-time nature of reserve service. Indeed, this has been identified by reservists themselves:

> R1: Most people get [the four weeks needed to undertake driving course] as holiday allowance for the whole year. So how can you go, 'Ah, fuck it, I'll take three months off and do it'?
>
> R2: You could smash it and probably have a great laugh, whereas if you've got a 9–5 [job], its pointless.[3]

This sentiment was frequently repeated:

> R1: I would say to anybody thinking of joining reserves, if you're in full-time employment, think about it carefully … Most people get 28 days holiday a year, and if you do join, you might do it for a year and give up all your time, but then after that, you might start getting a bit tired of giving up every day you have spare.
>
> R2: As soon as they turn it to 40 days –
>
> Mod: If they did.
>
> R2: – if they did, I'll be[leaving].[4]

These quotes highlight the increased burden on reservists' time as a result of FR20, and the limits of their willingness to accept this. Similarly, during the research it became clear that certain reservists, mainly the unemployed and self-employed, are able to commit the most time to the reserves. While this is not new in and of itself, the fact that they were then best positioned to take advantage of FR20's increased opportunities to attend courses and deploy– with the subsequent impact on career progression – was identified as creating a new imbalance in some sub-units between those in civilian employment (and who were the traditional backbone of the TA) and those who are not. Indeed, as a senior NCO in full-time employment elucidated: 'I'm disadvantaged to people that don't work. There's no way I can compete.'[5]

---

3   Interview 1.
4   Interview 10.
5   Interview 11.

The greater monetary benefits on offer and the impact of the recession further support the evidence of an emerging division between part- and full-time reserve service. Indeed, it appears that a core of reservists able to commit more time may be developing. As the reservists below remarked:

R1: Yeah, it's always the same people go away.

R2: At the moment I've actually stopped working so I can get my days in.[6]

Supporting these quotes, numerous sub-unit commanders identified that certain individuals, due to their ability to commit more time, were reaping the benefits of the increased opportunities, at the expense of some of their colleagues who could simply not spare the time to do likewise.[7] There were also reports of part-time reserve officers who had worked for a full month (un-deployed) due to the need for those in command appointments to work more closely with the regulars. Similarly, another reservist reported: 'We're getting less and less bounty hunters as well, who just turn up for the minimum for the bounty. They're kind of getting shipped out generally.'[8] Thus, it appears that one of the impacts of FR20 has been to begin to divide the Army Reserve between those who, primarily for reasons of time, are drawing closer to the regulars and with it increasing their professionalism, and those who cannot. It is perhaps not too much to argue that a split is therefore beginning to occur. While it is certainly true that the reserves have always catered to those with more time to commit to service than those with less, it appears this division has been accentuated by FR20's emphasis on courses and increased deployability. Indeed, it is arguable that the reserves are no longer a fully part-time organisation; some elements, especially its officer and SNCO corps, are being drawn into full-time reserve service. In its pursuit of a reserve integrated into the Whole Force, FR20 is thus beginning to challenge the traditional notion of what reserve service means by assimilating elements of the reserves into the regulars. Following King, it is possible to argue that this division is concentrating increasingly professional and full-time reservists into a core that is much closer to the army, while the traditional part-timers remain at the periphery. This threatens the traditionally distinctive nature of the reserves as a part-time volunteer organisation and could, conversely pose a threat to its long-term health as an institution. Indeed, there is evidence to suggest that already this is causing friction and some resistance. As one NCO reflected on the traditionally part-time nature of the reserves: 'We are never going to be the same as them, and we shouldn't expect to be.'[9] Given the need to maintain the reserves 'equilateral triangle', it remains to be seen if this trend is sustainable,

---

6   Interview 10.
7   Interview 11.
8   Interview 12.
9   Interview 2.

and what affect it will have in the long term. Recognising this fact, both Brazier and General Carter were aware of this danger, and have stressed the need for integration over assimilation, thereby recognising both the limits of the Whole Force and the distinctiveness of the reserves. Indeed, there are signs of a move away from other formerly central tenets of FR20.

SDSR 2015 and the Divisional Level as an Organisational Solution

Following Carter's October 2014 remarks indicating that the Army Reserve's role was 'for the worst-case' and that the message to potential recruits and employers would be 'refined', as discussed in Chapter 4, in February 2015 he elucidated on his apparent re-appraisal of FR20's original goals. At a speech at Chatham House, he outlined his position that: 'the obligation if you join it [the Army Reserve] is for training only ... we are not going to use it regularly and routinely, as perhaps was suggested a couple of years ago. Rather, it is there in the event of a national emergency.' He went on to state: 'That means it's much more straightforward, I think, for an individual to be a member of the Army Reserve ...'[10] At first glance, such a *volte face* represented a complete reversal of the intended role of the reserves to that envisaged in FR20. Indeed, taking Carter's October 2014 and February 2015 comments together, he appeared to be saying that the Army Reserve's role was much more similar to that of the old TA – a strategic reserve – rather than an operational one. This, of course, undermined another central tenet of FR20, and, initially at least, created further confusion as to what exactly FR20 is meant to achieve. Perhaps more significantly, it also seemed to be a recognition of the enduring organisational paradox that a more deployable reserve means a less recruited one.

However, the reality is rather more nuanced, and the 2015 SDSR is crucial to understanding Carter's position. Released in October, the 2015 NSS and SDSR differed markedly from its predecessor. Most notably, the threat from international military conflict was prioritised, while instability overseas, public health and natural disasters all moved into the Tier One threat bracket, having previously been Tier Two or unlisted in the 2010 SDSR. Overall, the 2015 SDSR increased the emphasis on the threat of a conventional war with a major power – predominantly in response to Russian aggression in the preceding five years – while also stressing the need for increased national resilience. Interestingly, the general erosion of international order and resulting chaos also made a more significant appearance, indicating a realisation that some of the stable planning assumptions underpinning Army2020 had been reconsidered.

The 2015 SDSR committed the army to be capable of quickly deploying a larger expeditionary force of 50,000 (compared with around 30,000 planned in FF2020)

---

10   General Sir Nick Carter, Comments made at Chatham House brief, 17 February 2015.

by 2025.[11] Crucially, it set out that the army would be expected to deploy a 'war-fighting division optimised for high intensity combat operations.' This division 'will draw on two armoured infantry brigades and two new[ly created] Strike Brigades to deliver a deployed division of three brigades'.[12] According to Carter, this top-down strategic re-orientation caused a change in the army's planning assumptions. Decisively, the SDSR's focus on the new quickly-deployable war-fighting division meant that the regulars would now provide the majority of its forces. This, of course, has major implications for the Army2020 readiness structure, based as it was on Reactive and Adaptive Forces of integrated regular and reserve components to deliver 'an enduring operation at medium scale in perpetuity.'[13] Moreover, the emphasis on resilience also requires that the regular army 'to be at higher readiness in greater numbers to deliver UK national resilience', such as flooding relief.[14] While the exact implications of this re-orientation for the Army2020 structure and readiness cycle are still being examined by the army under the 'Army2020 Refine' project, it is clear that at the highest level the focus on enduring operations and its supporting 'harmony guideline' is being replaced by the need to rapidly deploy greater mass on the battlefield and the principle that the nature of the task should be assessed and then the appropriate tour length and interval determined.

This fundamental shift away from some of Army2020's guiding principles toward Army2020 Refine is ongoing but it has already considerably affected the role of the reserves outlined in FR20. According to Carter, the army now intends to restructure the reserve into three echelons organised to support the army's new main goal of war fighting at the divisional level. Crucially, contrasting the regular and routine deployment of reservists, the reserves will now be tasked with providing the basis for reconstitution and regeneration of the regular army within this model, while also being available to support the regular's re-organisation for other tasks, such as an enduring operation or national resilience. Similarly, while this new design does not preclude the deployment of formed sub-units and units, it is no longer viewed as essential, as under FR20. Instead, the army has committed to routinely offer opportunities for reservists who are available to deploy, either collectively or as individuals, on operations and other tasks, when required. But decisively, according to Carter, reservists' 'can deploy on operations and exercise if they can spare the time, but their minimum obligation is for annual training now', thereby indicating a major shift away from FR20 that reflects an acknowledgement of the organisational difficulties of implementing the transformation.[15]

---

11 HM Government, *National Security Strategy and Strategic Defence and Security Review*, 29.
12 Ibid., 31.
13 General Sir Nick Carter, 'Opening Remarks', RUSI Land Warfare Conference, 28 June 2016, 2.
14 Interview, Carter, 11 May 2016.
15 Interview, Carter, 11 May 2016.

Three major categories of workable roles are now being considered for the reserves to reinforce the regular's re-orientation to the war-fighting division. This first category includes specialist units and individuals utilising their civilian skills, such as medical and intelligence experts. The second represents the majority of the Army Reserve and covers generalist combat, combat support, and combat service support units that will be trained collectively to a standard that is achievable from within an average allocation of 40 annual man training days. The last group consists of primarily specialist combat units such as the reserve parachute and special forces units, that due to their ethos and training, are available at higher readiness to deploy collectively. According to Carter, this more bespoke and flexible method of deploying the reserves better reflects the realities of their ability to recruit and train to full strength, and hence their readiness levels: 'It is a more plausible role and a more plausible narrative.'[16] However, in stating that the pattern of reserves deployment will be 'voluntary except at best effort', this indicates that, from an organisational perspective, Army Reserve mobilisation will be a hybrid of that of the traditional TA and that originally envisaged in FR20. Carter was instrumental in leading these changes, and since our interview they have been codified in the Army Reserve Sub-Strategy of 2017 around the 'Four Rs'principles: Reinforcement, Resilience, Regeneration, and Reconstruction.[17]

There has therefore been a clear and concerted move away from Army2020's and FR20's planning guideline that the reserves would be deployed routinely and regularly at the sub-unit level from roule four of an enduring operation. The fact that these two central tenets of FR20 have been revised indicates that they were not based on the organisational realities of the TA/Army Reserve. It also supports the findings in Chapter 6 of the major issues sub-units had recruiting to strength, the availability of equipment to train on, and ultimately delivering the 'hard' capability required of them outlined in Army2020 and FR20. Indeed, it appears that many of these grass-roots organisational issues were repeated across the Army Reserve, and that the high command was well aware of them, often from the outset. Nevertheless, due to the political origins of FR20, and the political appetite at the time to implement it, after first cautioning against the plan, the army's leadership did ultimately make a major effort to effectively implement FR20. In one respect, this recent policy revision therefore represents an acknowledgement that the army was right about the major difficulties associated with creating a more deployable reserve.

There are signs that FR20 has been revised due to other organisational frictions. The most prominent of these is, of course, recruitment. In July 2018 the Army Reserve's trained strength had risen to 26,790,[18] but again a large proportion of this increase is explained by changes to accounting metrics, in particular the October

---

16 Interview, Carter, 11 May 2016.
17 Mooney and Crackett, 'A Certain Reserve', 90.
18 UK Armed Forces Quarterly Personnel Report (July 2018).

2016 decision to include Phase 1-trained (i.e. basic trained) soldiers on the total trained strength of the army and the reserves.[19] This was justified as meeting a recommendation of the EST to rapidly increase trained strengths, and by reference to the fact that reservists and regulars do not need to be trained to Phase 2 to conduct numerous national resilience tasks.[20] It also reverted strength metrics to the previous system.[21] Nonetheless, as with other changes to recruitment metrics detailed in Chapter 4, while justifiable, such a move also had clear beneficial political and organisational outcomes by increasing the numbers on the books by over 2,400. Thus, taken together with the inclusion of full-time reserve service personnel, around 3,150 out of the reported growth of the Army Reserve is simply a result of changed metrics. In reality therefore, the AR has only grown in terms of real trained strength inflow by about 5,000 over five years. This is fairly shocking given the resources and effort spent during this time, and leaves the reserve well short of its 31,000 trained by April 2019. Moreover, in 2018, the rate of ex-flow from the Army Reserve began to rise above inflow, pointing to decreasing strength ahead. Indeed, Parliamentary questions informed by research in this book have revealed that only 14,920 army reservists received their annual bounty in 2017. The bounty is a strong indicator of fully trained and deployable strength and this suggests that the real strength of the active trained army reserve is much smaller than claimed.[22] Similarly, as identified in Chapter 5, there are also concerns regarding recruit quality. Interestingly, this and the difficulty reaching recruit targets has caused the NAO to recommend that recruitment targets should be heavily revised downward.[23] This represents an acknowledgement that the 30,100 trained strength by April 2019 target is unrealistic and unlikely to be achieved. Another of the major tenets of FR20 has been adjusted due to organisational friction.

A further example concerns budgets. FR20 pledged £1.2 billion to reinvigorate the Army Reserve between 2013–23; about £1.1 billion has been spent to date. Indeed, there is evidence to suggest that reserve training budgets have already been cut as the army prepares for further years of tight fiscal constraints. While a reduction in training budgets for the reserves can be justified by the argument that the newly integrated reserve force cannot be prioritised over other components of the Whole Force,[24] it does run counter to FR20's rationale. Here it is interesting to note the final analysis of the 2015 EST report:

---

19 See 'Changes to the definition of trained strength for the Army and resultant changes to the Ministry of Defence Armed Forces Personnel Statistics', available at https://www.gov.uk/government/uploads/system/uploads/attachment_data/file/537077/20160712-MOD_Personnel_Statistics-GTS_consultation_document.pdf, retrieved 21 September 2016.

20 Interview, Carter, 11 May 2016.

21 Connelly, 2018.

22 MoD response to written question from Madeline Moon MP.

23 National Audit Office, *Army2020*, 34.

24 Ibid.

> We are acutely aware of the current tautness the Defence budget, with significant risk in many programmes. Any further budgetary pressure, if realised, is likely to have a direct bearing on the ability to deliver FR20 – whether as a consequence of direct cuts to the programme or indirectly through reductions in activity which exacerbate recruiting and retention risk.[25]

It appears that these fears were well-founded. Spending cuts may already be responsible for falling recruitment, while, conceivably, reduced training activity could contribute to retention problems. Of course, as the EST's conclusion highlights, the blame for this can hardly be put on the army. But the fact that the army is confident enough to justify cutting the reserve budget hints that practical considerations may finally be trumping political ones, which in turn highlights the decreased political attention on the issue. It is noteworthy that after the fall of the Cameron government in June 2016, Brazier resigned as reserve minister in July, and later lost his seat in 2017. Indeed, the army leadership's management of the latter stages of FR20 displays a sensitivity to the transience of politics and an awareness that it would ultimately be left in control of the reserves again.

However, it is important to stress that these recent revisions do not mean FR20 has been abandoned. Far from it. Despite the need to 'manage resources efficiently' across the Whole Force, transformation of the Army Reserves is continuing with strong emphasis on collective training and opportunities to train and deploy with the regulars.[26] The system of pairing units – the army's own practical solution to increasing reserve capability – remains crucially important to both the regulars and the reserves and a central part of reserves policy. Complementing the evidence from logistics units on the increased availability of deployments, reserve combat and non-combat forces are being deployed much more frequently than in the past, predominantly in lower threat environments. For example, reserve infantry platoons and company groups have deployed to Cyprus, the Falklands and Ukraine recently, indicating the army's commitment to offering opportunities and deploying the reserves post-FR20. Moreover, reservists reported the availability of training deployments to Cyprus, Germany and Canada in a 12-month period, with three opportunities to deploy to the latter.[27] This marks a significant change from opportunities in the TA. Crucially, these deployments provide a means by which reserve professional standards should gradually increase, itself conducive to better mutual understanding and, ultimately, the respect of the regulars. This workable method of better integrating the reserves is also being complemented by other efforts that support FR20's political aim to reinvigorate the force and ensure its organisational survival by making it more capable. For example, young officer training has been made more flexible to better fit reservist circumstances,

---

25 CRFCA (2015) *Reserve Forces External Scrutiny Team Annual Report*, 3, London: NAO.
26 Interview, Carter, 11 May 2016.
27 Interview 12.

and new career paths that allow them to serve at regimental duty and on staffs introduced. Similarly, training for all ranks is being made more modular to fit the unique position of reservists.[28] Combined with these reforms, efforts to open up the coveted command appointments in regular units to reservists are being made, indicating a step-change in how the regular high command view the best reservists' contributions. Other measures are being made to introduce career management models that better reflect the realities of reserve service. Another signal of intent is the appointment of a reservist Major General to the Executive Committee of the Army Board, giving the reserve component more clout in terms of recognition, resourcing and policy making. Carter has stressed that the ultimate health of the Army Reserve will be determined by the re-establishment of a well-recruited, trained and vibrant officer corps. Taken together, these changes indicate that, supporting the findings in Chapter 6, some of FR20's most enduring successes will likely be in steadily changing the relationship between the reserves and the regulars, and with it the slow inculcation in the reserves of a more professional ethos based around that of the regulars but respective of the distinctions between the two. These cultural changes are likely to continue to be less obvious than the hard capabilities that the reserves were originally forecast to provide, but over time they may be equally as important. Nevertheless, apart from perhaps in the category three roles defined above, social cohesion is likely to remain more important in the reserves due to the different nature of its service and the fact that many explicitly join to experience the comradeship traditionally associated with army service.

Given the political battles surrounding its origins, and the criticism FR20 has received from numerous quarters, for his part Carter has understandably presented this revision of FR20 as primarily caused by the changed demands the 2015 SDSR placed upon the army, and in particular the re-emergence of the war-fighting division. In the context of chapters two and four respectively, this position can be seen as emphasising the strategic rationale for organisational reform. There is also an acknowledgement at the top of the army that the inability to recruit to strength was threatening the delivery of reserve capability to Army2020's roule four and beyond, thereby threatening the overall sustainability of the model. Problems delivering collective reserve capability compounded this, while paradoxically, the requirement to deploy collectively and more often increased the training burden on reservists and threatened to upset the balance between their service, employment and their families. Just as in the previous periods of reserve reform, major organisational challenges, many rooted in the very nature of reserve service itself, not only prevented FR20 from reaching two of its primary goals, but these goals in and of themselves in turn threatened the overall sustainability of the reserve organisation. Due to these organisational

---

28 Interview, Carter, 11 May 2016.

difficulties, and more importantly, the political sensitivity and media interest surrounding FR20's success, the new emphasis on the centrality of the war-fighting division and its implications for Army2020 is fully understandable. But it may only partially explain the recent revision of the reserves policy.

Unlike its predecessor, the most recent SDSR has been widely praised as a comparably strategically sound document that seeks to align ends, ways, and means.[29] However, while the Army Reserve issue is very unlikely to have influenced the increased emphasis on major conventional conflict outlined in the SDSR (this was based on the National Security Risk Assessment), it is possible to argue that the change in emphasis that accompanied it presented the army with a relatively fortuitous opportunity to address the major organisational problems FR20 had created. Indeed, given the political infighting, intra-service rivalry and recurring organisational frictions that FR20 had caused, the 2015 SDSR's emphasis on the war-fighting division offered the army an organisational solution by which to extricate itself from the transformation's most ambitious – and clearly unworkable – elements. Most importantly, this solution was based on strategic rationale and provided a perfectively justifiable narrative given the changes occurring to the army's role and structure. Crucially, the timing and content of Carter's February 2015 speech, in which he spoke first of the need to fight at the divisional level and then directly followed this with his thoughts on the changed role for reservists, hints that this organisational solution may have been understood during SDSR's planning phase. That is not to suggest that the army's emphasis on the division was specifically designed in order to organisationally extricate itself from the failing elements of FR20 – it clearly wasn't. Edmunds has detailed the transnational nature of British defence, and with the centrality of divisional level operations being adopted in the US for similar strategic reasons, Britain knew it would need to follow suit to retain both political clout and interoperability,[30] but there was arguably an awareness within the army of the political and organisational benefits that such a change in its operational posture would have on FR20. This is likely to have been complemented by the growing doubt that the Army Reserve could contribute the required capability to the later roules of a deployment. By June 2016, in his first public announcement that the rationale underpinning FR20 had changed, Carter stated that 'one of the advantages of these new Defence Planning Assumptions [DPA] is it allows us to think more from first principles about what the role of the Army Reserve should be.'[31] He went on to clarify:

---

29 International Institute for Strategic Studies (2015) 'UK augments military and counter-terrorism capacities'. *The Daily Telegraph*, 23 November 2015, 'SDSR: Lord Dannatt's reaction'.

30 Edmunds, T. (2010) 'The Defence Dilemma in Britain', *International Affairs*, 86(2); King, A. (forthcoming) *Command.*

31 Carter, 'Opening Remarks', 3.

You recall that a year ago, given the [DPA], [the Army Reserve] was there very much to backfill and integrate a regular structure which was designed to manage [an] enduring operation in perpetuity. Now it is there for reconstitution and regeneration. It is there in the event of a nationally recognised emergency. Now that's not to say that the reservists are not able to take their part if they can afford the time and effort to be able to deploy alongside the regular components, but they are there in true obligation terms for the worst-case. Now that is proving to be easier to recruit for.[32]

Carter's final sentence is particularly interesting. It is therefore not too much to suggest that the SDSR, perhaps more by implication rather than design, presented a solution to the major organisational frictions FR20 had caused the reserves by calling for their regular and routine deployment at the sub-unit level. The 2015 SDSR, and the rapidly deployable war-fighting division it called for, has allowed the army to extricate itself from the most problematic parts of FR20. Crucially, given its political origins, it has allowed this to occur with little fanfare or political cost and has and is being explained by an altered narrative that can be justified by changed strategic circumstance.

Indeed, following this argument and building on that in chapters four and five, it appears that less than two years after it was unveiled, in 2015 another key moment in FR20's evolution had occurred. This evolution was again heavily influenced by strategy, recruitment and organisational friction; the SDSR's new vision for deploying an army division provided a relatively unique chance to solve FR20's organisational shortcomings. The content and timing of Carter's February 2015 Chatham House speech indicates that the two issues had been linked at this time. Similarly, the timing of Carter's October 2014 remarks on the 'refinement' of the message to reservists about their regular and routine deployment, appears significant for three reasons. Firstly, coming soon after he had become CGS, and hence untainted by previous allegations made against the army's senior leadership of wanting FR20 to fail for their own organisational survival reasons, Carter's remarks can be seen as highlighting that a politically neutral re-appraisal of FR20 indicated it was failing in some critical areas. Secondly, his October 2014 remarks were likely something of a political litmus test, allowing Carter to gauge senior political commitment to FR20 after the heavy weights had divested their capital. Thirdly, and most importantly, this public re-appraisal was supported by practical facts, in particular coming just before new figures would again highlight an inability to recruit to strength, thereby seriously undermining of increased reserve deployability and highlighting the need to keep 'balance' in the reserves.[33] This is likely to have been complemented by further indications coming up the chain of command of the organisational frictions caused by the focus on recruitment

---

32 Ibid.
33 Carter, Chatham House, February 2015.

activity and re-roling at the sub-unit level. Thus, the reserves recruitment issue was constraining its ability to provide the organisational output required by Army2020's deployment cycle, and thereby threatening the overall coherence of the plan. Mirroring the lessons of the past, recruitment and other organisational issues interacted with strategic considerations to shape the transformation of the reserves at this moment.

Moreover, as with the past periods of reform, while the issues influencing the direction of the reserves have been remarkably constant, the organisational solutions adopted to address them have been a product of their own time. The importance of the 2015 SDSR and the re-emergence of the war-fighting division supports this argument. Similarly, without the major political input and the context of organisational survival that shaped FR20, it is arguable that in the most recent revision of the transformation, senior leaders within the army had much more scope to alter course. It appears that another key moment in the evolution of the reserves has occurred, one that has returned its operational role to much closer to that of the TA than originally envisaged, but with a much higher degree of integration with the regulars. This arguably reflects a more realistic assessment of the Army Reserve's organisational nature and what it can realistically provide in terms of capability. It also indicates an awareness that the fundamentally political goal of arresting the neglect of the reserves has been met, and that a differentiation between the need for this and the problems associated with outsourcing operational capability to the reserves was due.

Finally, the re-emergence of the war-fighting division has interesting implications for the future of the post-Fordist approach to military organisation. Indeed, given the similar changes afoot in the US, it suggests that mass is seen as increasingly important to future military operations. It also indicates a realisation of the limits of the effectiveness (and indeed long-term efficiency) of outsourcing to the reserves, and a desire to maintain core capabilities within a larger regular army formation whose primary role will be fighting conventional or near conventional wars. Interestingly, this re-orientation of the army under Carter to an emphasis on mass provides evidence of a tacit awareness of the limits of the 'Total Cost' Whole Force approach to organising military forces. While it would be wrong to suggest that the re-emergence of the war-fighting division means an end to the four processes of post-Fordist military organisation – it clearly doesn't – what it does provide is further evidence of the army's realisation that the stable assumptions underpinning rotational readiness and deployment structures are not best designed to address strategic shocks. The growing risk of a major conventional war, other strategic shocks, and even the decline of international order into chaos, requires both greater mass and organisational flexibility. Changing perceptions of the nature and scope of threat have driven this return to the rapidly deployable division, but it may also mark the start of a re-assessment of the post-Fordist military processes and structures that have delivered efficiencies but simultaneously reduced strategic flexibility. Indeed, Carter's statement that the reserves' primary role has returned to providing a strategic reserve for 'a national emergency' echoes Lamb's about the reserves' real utility in providing cheap, scalable mass in such an event. Nevertheless, as Chapter 4 showed, in the longer term, removing the reserves from the army's readiness cycle could make them more vulnerable to defence cuts again when the political winds change.

FR20 as a Transformation?

The centrality of FR20's original goal of the transformation of the reserves is difficult to deny. The Independent Commission submitted its proposals under the banner of 'transforming the reserves' and mentioned reserve transformation a further eight times, calling for a 'reinvigorated Reserve transformed into an integral component of the Whole Force.'[34] Although he had referred to reserve transformation numerous times during the consultation process,[35] when introducing FR20 in July 2013, Hammond – likely cautious about over promising and under-delivering – stated that the new policy aimed to 'revitalise' reserve forces rather than explicitly transform them. It will be remembered from Chapter 1 that the document itself clearly stated that 'FR20 is part of the wider Transforming Defence campaign that is aiming to transform our Armed Forces and deliver Future Force 2020.'[36] It also specifically mentioned reserve transformation a further three times. Clearly, the army and the wider defence establishment viewed FR20 as a transformative process.[37]

The question is then, has FR20 transformed, or is it transforming, the reserves? Of course, much depends on the definition of transformation. As Foley et al. have argued, military transformation is in fact simply another name for innovation.[38] Farrell has also distinguished between top-down innovation bottom-up adaptation undertaken in response to operational pressures. It will be remembered that Grissom has argued that for an innovation to be recognised in the academic literature it has to meet three criteria. Firstly, 'an innovation changes the manner in which formations function in the field'; i.e their operational praxis. Secondly, the innovation must be significant in scope and impact, a definition that Grissom recognises implies a consequentialist understanding. Finally, 'innovation is tacitly equated with greater

34 *The Independent Commission*, 38.
35 'Consultation launched on the future of Britain's Reserve Forces' available at https://www.gov.uk/government/news/consultation-launched-on-the-future-of-britains-reserve-forces, retrieved 28 July 2016.
36 *Future Reserves 2020*, 59.
37 *Transforming the British Army*; *Transforming the British Army: An Update.*
38 Foley et al., 253.

military effectiveness.'[39] FR20's original goal of creating a better trained and equipped Army Reserve, held at higher readiness deploying routinely at the unit and sub-unit level, in order to increase the army's capability, appears to meet all three of these criteria. The unit and sub-unit aims sought to change the way the TA had predominantly been used in the field. FR20 itself stated that so profound was the cumulative effect of this policy in terms of scope and impact that it represented a transformation. And tying the more deployable reserve into the Army2020 readiness cycle was designed to increase overall military effectiveness and efficiency. FR20 was therefore clearly a transformative attempt to turn the Army Reserve into an operational rather than a strategic reserve. So has it succeeded?

Most of the top-down transformation literature is based on archival research of past periods of military change. It is therefore important to note that FR20 is still ongoing and that I have used predominantly recent and current data to examine the policy. There is small potential for its trajectory to change again. Nevertheless, this research has revealed that the central tenets of FR20 have been revised downwards since the policy was introduced. Most importantly, there has been a major revision of the reserves role detailed in FR20. The routine and regular deployment of reservists on operations appears unlikely to happen. Instead they will be used for a 'worst-case' scenario. As such, the fully operational role of the reserves has been modified; although likely to be more capable, overall, the model of reserve deployment on operations will remain closer to the strategic reserve role of the TA. Similarly, there has been a less obvious but equally profound move away from the deployment of reserve sub-unit and unit formations. This has been driven in part by the organisational reality that many reserve sub-units could not have provided the required capability in perpetuity anyway. As I have shown, this bottom-up organisational resistance in turn undermined the overall coherence of Army2020's rotational deployment plan, adding another reason for a revision of reserves policy. The effect of a reduced reserves training budget in the years up to the end of FR20 has compounded this. Indeed, given these recent revisions, it seems clear that there will not be a major pan-organisational change in how the Army Reserve is used in the field.

The evidence is perhaps less clear concerning the scope and impact of FR20. On the one hand, despite major investments in training, recruitment and equipment, closer integration with the regulars, many more opportunities for reservists, and relatively high levels of reservist satisfaction (74 per cent),[40] the requirement for the majority of the Army Reserve to contribute to the army's operational effectiveness in the manner FR20 detailed is not being pursued as originally envisaged. On the other hand, these factors are contributing to significant changes within the Army Reserve that could increase its military effectiveness in the long term. Following

---

39 Grissom, 'The Future of Military Innovation Studies', 907.
40 MoD (2018) Reserves Continuous Attitude Survey, 8.

Kier and Farrell, perhaps the most important of these changes have been cultural-normative, with a greater sense of professional ethos and professional pride emerging that emulates that of the regulars. This is supported by the decline of the importance of the traditional drinking club and an increasing perception of reserve service as a 'job'. Through its continued commitment to pairing and integration with the regulars, and greater availability of courses and deployments, this cultural shift is likely to maintain momentum, in the short to medium term at least. Despite the political de-investment at the top level of government (which has continued under Theresa May's premiership), for the moment the army appears committed to revitalising the reserves, at least through these processes. Therefore, it is possible to argue that a major cultural change is underway in the Army Reserve, and that by encouraging professional standards, this is gradually affecting the operational praxis of the organisation as a whole. While the wider impact of professional values is different to the cultural emulations discussed in the innovation literature, King and others have already argued it to be central to the recent transformation of Western European armed forces' effectiveness and that of the combat infantry in particular.[41] As I have shown, given the distinct organisational nature of the reserves, and the fact that reserve cohesion is still based on social bonds, inherent bottom-up, micro-level factors can limit the ability of military organisations to emulate those they wish to. Thus, transformation is likely to take longer in the reserves and may not be implemented as fully, but nevertheless it should ultimately increase its military effectiveness. It appears that one of the criteria for military innovation has indeed been met. FR20 has been a partially successful transformation.

What does this partial success tell us about the academic literature on transformation? Most obviously, this study has addressed a major gap in the innovation literature by examining reserve, and non-combat, forces for the first time. It has also utilised the wider sociological literature on post-Fordism, professionalism and cohesion to more deeply understand the nature of the processes at play than in the transformation literature. This approach has revealed that, as might be expected given the traditionally part-time nature of reserve forces and these organisation's more limited ability to demand its members' time, there are often more deeply ingrained organisational factors resisting change than in regular forces. In the case of the Army Reserve, some of these are related to the organisational paradoxes identified in the TA by Walker, and more recently articulated by reference to the 'equilateral triangle' by Wall and Carter. For example, the fact that increased training and deployment demands can negatively impact reserve turnout, recruitment, and retention, delineates a major difference with the regulars. Moreover, as reserve officers and indeed SNCOs are generally less experienced, and have less time for management and administration tasks, both top-down and bottom-up transformative processes take longer to effect change.

---

41 King, *The Transformation of Europe's Armed Forces*; *The Combat Soldier.*

Supporting Chapter 2, the nature of reserve organisation and service therefore in and of itself makes transformation inherently more difficult than in regular forces.

This study also contrasts the consequentialist, even positivist, nature of the majority the military transformation literature. It helps reverse that trend by examining an uncompleted transformation, and within this, how external top-down political direction failed to produce the organisational changes originally envisaged. As I have shown, political elites' plans for the reserves were not grounded in organisational reality but in intra-party politics. The real impetus behind FR20 came from a coalition of both senior and junior politicians with considerable leverage over a new and weak Prime Minister, who were equipped with an admirably strong desire to end the neglect of the reserves that they perceived would continue if left to the army's high command. However, these intensely intra-party political origins also resulted in severe intra-service rivalry and the army's initial strong resistance to FR20. Most crucially, as I and official reports have shown, the cumulative effect of these origins resulted in an initially ad hoc and poorly modelled plan far removed from the organisational realities of the Army Reserve. FR20 was therefore adjusted at each step in its development due to army resistance and organisational friction in the reserve. Once implemented, these frictions ultimately undermined attempts to transform the reserve into an operational force. Bottom-up resistance therefore severely limited the impact of central tenets of FR20. Complementing both Catignani's and Harkness and Kunzerb's works, the case of FR20 and the Army Reserves shows how low-level organisational resistance can curtail top-down politically-imposed innovation. It also neatly supports Allison and Kaufman's arguments that broader institutional change is driven by elites, revised by stakeholder resistance and organisational friction, and ultimately results in a re-booted version of the organisation. Indeed, this perfectly surmises the origins, evolution and impact of FR20.

FR20 and Civil–Military Relations

The story of FR20 also raises interesting implications about British civil–military relations during this period, which are worthy of brief discussion here. In *The Soldier and the State*, Huntington described and indeed called for the objective civilian control of the military as the most effective way of maintain the armed forces' capability while remaining relatively de-politicised from party agendas. For Huntington, professional military service should be removed from politics and subservient to the state.[42] In return for this, government should not interfere in military matters.[43] However, Janowitz outlined a more subjective model of civil–military relations based on his view that the large technocratic and bureaucratic

---

42 Huntington, *The Soldier and the State*, 79, 466.
43 Ibid., 83.

US military of the 1950s increasingly resembled a modern civilian corporation. As a result, he argued that the military was coming under increasing subjective control as it became more reflective of civilian society as a whole; it therefore needed to be close to government, politics and society to reflect it.[44] Much more recently, Peter Feaver has convincingly argued that civil–military relations in fact resemble a principle agent theory 'game' of strategic interaction between civilian leaders and military agents. In this model, civilian leadership controls the military through monitoring and punishment, and the military can either 'work or shirk' based on its expectations of punishment.[45] Regarding British civil–military relations, Egnell has argued political control is highly centralised, enabling political control of the military to be conducted with low political costs. Because of this 'low cost of monitoring' the armed forces, military officers, if mindful of their careers, have always had to stay in tune with the wishes of the political leaders.'[46] For this reason, Egnell argues that British civil–military relations more closely resemble the 'Janowitzean' model. Edmunds has also noted the complexity of British defence policy space, with deep inter- and intra-service divisions over the role and resourcing of the armed forces and the wider impact of political and economic interests.[47]

The political origins of FR20 and the army leadership's initial reluctance to instigate the reforms supports Egnell's view that British civil–military relations are closer to Janowitz's model than Huntington's. Senior and even junior politicians were closely involved in shaping and directing the new reserve policy, often against the wishes of the army, who themselves at times were willing to attempt to mobilise public support for their position through comments and leaked reports to the media. Thus, far from being detached from politics, the army's leadership was aware of the need to be politically savvy in their arguments and narratives to ensure their organisation's survival. Clearly, both Dannatt and Wall had major disagreements with their political masters in the defence ministry and wider government over the cuts to the army and the reserve plan in particular, indicating the willingness of the army's leadership to 'stand-up' for its interests against politicians intent on overstretching the army in Iraq and Afghanistan, and reducing its size and capability in the drive for efficiencies, respectively.

Indeed, the case of Dannatt is of particular interest as it was widely acknowledged that his poor relationship with Labour Prime Ministers Tony Blair and Gordon Brown, and the frankness with which he articulated the army's interests vis-à-vis government policy during the Iraq and Afghanistan wars, ultimately cost him

---

44 Janowitz, *The Professional Soldier*, 420.
45 Feaver, P. (2003) *Armed Servants: Agency, Oversight, and Civil–Military Relations*, Cambridge, MA: Harvard University Press.
46 Egnell, 'Explaining US and British Performance in Complex Expeditionary Operations', 29(6).
47 Edmunds, 'The Defence Dilemma in Britain', 382–88.

promotion to CDS.[48] The Labour Government clearly viewed Dannatt as too political. Such an outcome, and the fact that some in the MoD later believed that the army's senior leadership wanted FR20 to fail, supports Feaver's view of the principle agent 'game' where both sides have leverage over the other in terms of determining careers and supporting policies. Indeed, it is interesting to note that Haughton, a supporter of FR20, was later somewhat surprisingly appointed CDS having not led his service, as is tradition. Similarly, as the principal architect of Army2020, Carter would have done his chances of promotion no harm either as he became CGS and later CDS. Within this 'game' context, the role of personality in British civil–military relations appears important, given the small size of senior circles in the defence community. Following the poor relations Dannatt had with the Labour government, his successor, Richards, and Brown were keen to cultivate better relations after his departure.[49] While personalities are clearly important, a change of personnel at senior level also often provides an opportunity for one or both sides of the British civil-military divide to re-assess their relationship and address them.

While not explicitly about British civil–military relations, *British Generals* provided an interesting insight into the nature of predominantly army-government relations during and after Blair's premiership. Numerous generals, including Lamb, outlined that the lack of ability of government to clearly articulate its strategy undermined the effectiveness of operations in Iraq and Afghanistan.[50] Meanwhile, Geraint Hughes has noted Blair's interventionist belief that the UK had the manpower and resources to act as a global police force alongside the US was 'absurdly grandiose'.[51] There was therefore a failure to politically support operations with a coherent and achievable strategic vision and the resources to enable this.[52] Despite the strains that these operations put upon the relationship between the army's leading generals and the rest of government during this period, Lamb, and Strachan, have commented upon the lack of resignations amongst the former.[53] For organisational, professional, and personal reasons generals remained in post. Similarly, former MoD senior civil servant Desmond Bowen noted how the relationship between ministers and generals was an unbalanced one, with the latter clearly subject matter experts 'not beyond threatening that they will expose the fact that military advice is turned down, if that course is not accepted.'[54] Clearly

48 *The Daily Telegraph*, 17 July 2009, 'General Sir Richard Dannatt: profile'.
49 Richards, *Taking Command*, 297.
50 Lt Gen. Graeme Lamb (Ret.), 'On generals and generalship', in Bailey, J., Iron, R. and Strachan, H. (eds) (2013) *British Generals in Blair's Wars*, London: Routledge.
51 Hughes, G. (2013) 'British Generals in Blair's Wars: A Review Article', *The Round Table*, 102(6).
52 Cawkwell, T. (2015) *UK Communication Strategies for Afghanistan, 2001–2014*, Farnham: Ashgate Publishing.
53 Strachan, H. 'Conclusion'; Lamb, 'On generals and generalship', 153–56, in *British Generals*.
54 Bowen, 'Political-military relations', *British Generals*, 277–78.

then, both generals and politicians were capable of playing politics. The evidence presented in *British Generals* therefore further supports both Feaver's principle agent 'game' model and Egnell's argument that British civil–military relations follow the Janowitzean model with closely integrated military bureaucracies competing with other government entities in ways that mirror civilian society.

Nevertheless, in as much as it directly addresses British civil–military relations, the context for *British Generals* is a sustained period of war-fighting and stabilisation operations. As I have shown, the drive for peacetime efficiencies and the government's political desire to reinvigorate the Army Reserve and integrate it into the army's deployment schedule also caused major strains between politicians and army generals. These were based on peacetime issues of global strategic vision, force structures, funding, and the organisational realities of transforming the reserves, rather than operational pressures. Crucially, the trend of strained army-government relations continued under the Conservative government; only once Carter became CGS was there an apparent reversal, supporting my argument about the importance of both personality and personnel change in recent British civil–military relations. Indeed, despite the traditional view that the Conservative Party is more favourable to the military and 'strong on defence', primarily due to its political ideology concerning state spending, it has overseen the reduction of the British Army to its smallest size since the Battle of Waterloo. While size is not in and of itself an indicator of military capability, the fact that the most effective army in the world, the US Army, and its British counterpart, has recently re-discovered the importance of mass indicates that this may not have been the wisest, or most cost-effective, policy in the longer-term.

Finally, what can the army learn from the experience of FR20? This is a question that numerous senior officers have asked. On the one hand, the army's senior leadership is legally and morally bound to take the direction of its political masters, even if the policy that follows this direction is flawed in places. In reality, senior officers have only two options when faced with such direction, either implement the policy (wholeheartedly or less so) or resign. In *British Generals* Lamb denounced the failure of generals who had been found wanting on operations in Iraq to resign.[55] While the situation concerning FR20 was different in that the operational competence of generals was not being tested, it is noticeable that while numerous senior officers resigned over the cuts to the army which underpinned FR20, no head of service did so.[56] This raises the question of the politicisation of the most senior appointments within the British defence establishment. Dannatt was accused of 'playing politics' during his tenure as CGS, while similar accusations were made against Wall and some of his team in terms of wanting

---

55 Lamb, 'On generals and generalship', 146.

56 *The Daily Telegraph*, 30 June 2012, 'Army's most senior female officer quits amid cuts anger'; *The Daily Telegraph*, 2 July 2012, 'At least six "talented" generals quit Army over defence cuts'.

and allowing FR20 to fail, although both strongly deny such accusations. Of course, resigning at the pinnacle of a 30-year career is a difficult decision, not to be taken lightly. But by not resigning over points of policy they strongly viewed as detrimental to their respective organisations, and indeed national interest, service chiefs essentially passed responsibility for delivering flawed programmes down the chain of command. As this research has shown, this has forced some sub-units to play politics themselves, increasing politicisation at lower organisational levels, a development that most appear uncomfortable about. However, more junior officers have less agency to reject such policies already accepted by the army's leadership as their careers depend on delivering the mission set for them by the chain of command. Numerous interviewees cited that one impact of FR20 has been to increase inherent tensions between sub-units who remain dubious about some of the changes, and higher commands that are responsible for implementing them. Indeed, one question the army asked to be included in the group interviews was the degree to which sub-unit personnel trusted the army's senior leadership in respect to FR20, indicating their awareness of the problem. The failure to reject poor policy at the top can result in greater politicisation and organisational friction down the chain of command. While service chiefs no doubt find themselves in a difficult position due to the political consequences of resigning, in order to ultimately protect the organisation from flawed policies, and highlight their inadequacies (not to mention the strong convictions of the chiefs that would likely win respect through the chain of command) perhaps this action could be taken more frequently. It is very interesting to note that while a number of army chiefs have been replaced, only one has resigned; Sir John French in 1914, as a result of the Curragh mutiny. There is a culture in Britain of army chiefs continuing to serve despite major disputes with their political masters. This has become increasingly prominent since the defence cuts of 2010 and the release of the Chilcot report in July 2016. While this culture is embodied by an awareness that high rank bears a responsibility to the political system and to the nation (and can also be justified by the fact that changes would be pushed through by a successor anyway), as this research has identified, such a culture is also clearly not without its risks. Similarly, given its rarity, any chief's resignation could have a major impact on civil–military relations in the wake of their departure. Nevertheless, the lack of resignations suggests an unwillingness at the highest levels of the army's leadership to take a career-ending stand over politically-imposed policies, parts of which were known to be unworkable.

There are other, less drastic lessons the army can learn. One major lesson is that the TA and Army Reserve have drawn, and will continue to draw on, strong political support external to, and at times in spite of, the regulars. The lesson of history is that the army therefore prioritises itself and neglects its reserve to its longer-term peril. Understanding the political importance of a capable army reserve should be central to future senior commanders, and would be ultimately beneficial to the regulars, reservists and wider British civil–military relations. Conversely, if service chiefs are going to play politics, they may need to get better at it. General Carter has provided a masterclass here. In terms of future reserve transformations, the army would also do well to better understand the organisational realities of reserve service. A key organisational paradox that, it must be said, most of the army leadership recognised from the outset, is that outside of a national emergency, a more deployable reserve means a less recruited one. While acknowledging that FR20 was politically imposed and implemented rapidly to reduce spending, future reserve transformations would benefit from a longer change management consultation period and more coherent, yet sensitive, direction from the army. Indeed, aside from the army leadership's reluctance to implement what it judged an unworkable plan, much of the difficulty in implementing successful transformation in the reserves has been caused by insufficient sensitivities to the different organisational nature of reserve service and the intricacies this creates. A final lesson learnt concerns the strategic messaging behind FR20. As this research has shown, the cutting of the army and the growth of the reserves were fused in an ad hoc fashion under Fox. Indeed, Fox himself was later a strong advocate for not cutting the army until the reserves were fully manned. But allowing the revitalisation of the reserves to be portrayed as simultaneous compensation for a vastly reduced army strength represented another mistake by military leaders, and indeed their political masters. As Carter has noted 'it is disappointing that the recent debate about the importance of the Army Reserve has too often been confused by the conflation of the regular army and the growth in the Army Reserve'.[57] In future, more careful coordination of transformative processes, and the messaging accompanying these, would be beneficial to both the regulars and the reserves.

FR20 and Society

That FR20 aimed for a transformation of the reserves has been proven. But it has perhaps been forgotten all too easily that it also aimed for a transformation of British society. Unveiling FR20, Hammond announced that 'Above all we [the government] seek a new relationship with society.'[58] Although all mention of 'Big Society' had been dropped since appearing in the Independent Commission, FR20 was replete with references to its attempt to change British society's attitude to reserve service. Across its pages, it spoke of 'harnessing the volunteer ethos of society to tap into the best talent the country has to offer' while arguing that 'greater reliance on the Reserves is more cost-effective for the nation' but also requires 'a greater willingness by society as a whole to support and encourage

57 Carter, General N. (2014) 'Army2020 The Army Reserve', *Army Reserve Quarterly*, Autumn, 12.

58 *Future Reserves 2020*, 7.

reserve service.'[59] Decisively, it stated: 'What we are asking is significant and it will require a cultural shift both in society as a whole and within the Armed Forces. This won't happen overnight; it will take time to achieve.'[60] Clearly, FR20 aimed to transform the nature of society's relationship with the reserve.

In terms of the British military, FR20 indicated a departure from the focus on the core professional force that had become one of the most defining characteristics of the British Army. The nature of this shift had previously been heralded in *The Times*, where, under a headline announcing 'The day of the "citizen soldier" has arrived', Wall noted that Britain had depended on the commitment of its citizen soldiers 'for generations' and called for society to support the growth of the reserves.[61] Interestingly, however, Wall's use of the term distorted the original meaning of citizen-soldier. Cohen has defined this as the distinctive motivations of soldiers, their representativeness of wider society, and their primarily civilian identity in the conscript US military of the 'Great[est] Generation'. In contrast, their professional successors are volunteers, increasingly unrepresentative of society, and have military rather than civilian identities.[62] Thus, Cohen concluded that the twilight of the citizen-soldier was nigh, and it had been caused by the drive for a professional military. In *The Combat Soldier* King adds much evidence to Cohen's claims, indicating that there has been a profound change in US and British combat forces caused by professionalism. Thus, in calling for the rise of the citizen-soldier in society, Wall, and FR20, were emphasising an aspect of a past model of military service without acknowledging the distinctive nature of history and society that had shaped these military forces in the past. Moreover, both appeared to forget that, as I detailed in Chapter 2, the past suggested that increased civilian involvement in the military occurred during periods of high threat, and even then conscription was often resorted to.

There is much evidence that society itself has profoundly changed from that which gave us the Great Generation. In *Bowling Alone*, Robert Putnam details the decline of community in America, which he describes in terms of 'social capital'. This refers the common bonds of reciprocity between citizens, which, crucially, are independent of market forces. Putnam argues that due to historical and societal factors – most prominently the Depression, the New Deal and the Second World War – the sense of civic duty was the highest amongst the Great Generation born between 1920–40, and has been declining since.[63] To back his claims, Putman cites evidence of declining voting rates, voluntary organisation membership, and sports

---

59 Ibid., 10, 8, 13.

60 *Future Reserves 2020*, 9.

61 *The Times*, 24 February 2012, General Sir Peter Wall: 'The day of the "citizen soldier" has arrived'.

62 Cohen, 'Twilight of the Citizen-Soldier'.

63 Putnam, R. (2000) *Bowling Alone: The Collapse and Revival of American Community*, London: Simon & Schuster, 357.

playing, arguing that the individualism of the 'baby-boom' and 'X' generations which followed them has led to a major, and potentially terminal, decline of community in the US from the mid-1960s onwards.[64] Ronald Inglehart's works, *The Silent Revolution*, and *Cultural Shift in Advanced Industrial Society* have provided further quantitative evidence of a decline in collective identities related to the state, public life (including religion) and employment, and the rise of more individualistic and pluralistic Western societies.[65] Interestingly, King has used Putnam's work to argue that Western society as a whole is professionalising, and that its militaries are simply reflective of this wider change. Indeed, the recent attempt to grow and transform the Army Reserve, sitting as it does between society and the increasingly detached professional military, appears to be a particularly useful paradigm within which to explore modern British society.

As has previously been discussed, there was much confidence amongst those calling for its revitalisation that the Army Reserve could quickly be recruited to strength. Based largely on the observation that Australia could muster a reserve force of 19,000 from a population of about a third the size of the UK's, the Independent Commission stated that an Army Reserve target strength of 30,000 by 2015 was achievable. FR20 revised this date to 2018 (later April 2019), and stated confidently: 'The total requirement presents only 0.15 per cent of the overall UK workforce and, in an historic context, we require only about half the strength of the Reserves as they were in 1990.'[66] With the other reserve forces removed, as at May 2013, the new Army Reserve would therefore represent 0.10 per cent of the workforce.[67] On the face of it, this appeared justifiable: the Options for Change programme had reduced the TA's establishment from 76,000 to 63,500, the latter representing a much higher 0.24 per cent of the workforce at the time.[68] However, if the Independent Commission had conducted further historical analysis it may have been less confident in its target strengths. Analysis of the mid-1960s – precisely where Putnam identified the beginning of a shift in societal values in the US – reveals much greater TA participation rates. Before the Carver-Hackett reforms of 1964, the TA's trained strength was 107,500, or 0.43 per cent of the workforce at the time.[69] The 2013 workforce was almost 5 million more than those in 1965 and 1990, but even when this is accounted for, between 1964–2013 there has been a 76 per cent decrease in Army Reserve trained strength relative to the workforce over this period. While the workforce metric is limited by the higher numbers of females

64 Putnam, *Bowling Alone*, 31–32, 259.
65 Inglehart, R. (1977) *The Silent Revolution*, Princeton: Princeton University Press; Inglehart, R. (1990) *Culture Shift in Advanced Industrial Society*, Princeton: Princeton University Press.
66 *Future Reserves 2020*, 14.
67 Office of National Statistics (2013) 'Labour Market Statistics, May 2013'.
68 Lindsay, C. (2003) 'A Century of Labour Market Change', Report for Office of National Statistics, 136.
69 Ibid.

working since 1990, this limitation is offset by the fact that the UK's population grew 16 per cent between 1965–2015. This suggests that if society had remained the same as in 1965, increasing participation rates in the TA would have been very easy. I have already noted Caddick-Adams' argument about the TA's consistent inability to recruit to full strength. More broadly, Strachan has noted that the British military has always struggled to recruit volunteers, and also the decline of militarism in British society since the end of conscription.[70] Edmunds et al. have also correctly identified that Britain's army and reserve army recruitment has historically been closely related to the public's perception of external threat.[71] Indeed, one major explanation for the recent difficulties both forces have experienced recruiting to strength is the negative impact of the wars in Iraq and Afghanistan, which have caused greater public distrust of their political leaders and questioning of the utility of force in general.[72] As Houghton has remarked:

> … rightly or wrongly, the legacy of the conflict in Iraq and Afghanistan have been, and still are, hugely challenging. They have affected some people's perception of the beneficial utility of Armed Force, of the competence of [the British] Defence [establishment] and the wisdom of past government.[73]

Thus, the shadow of post 9/11 interventions is very important for understanding the current context. But it also seems clear that difficulty recruiting the army and its reserve to their newer, historically small establishment, indicates that there has been a major change in the nature of British society since the 1960s as well.

Indeed, Edmunds et al. have identified that some of the recruitment problems are due to changes in British society, and have labelled FR20's recruitment goals 'over-optimistic, and perhaps even naïve' for failing to take stock of these.[74] In a relatively brief discussion, they draw on data from recent British Social Attitudes Surveys to argue that the British population born after 1979 in particular have increasingly liberal views and a greater preference for individual over collective identities.[75] Citing evidence of a decline in religious, political, and trade union activity, they argue that this group – predominantly Generation Y or 'Millennials'– are more sceptical of collective endeavours and 'suspicious of the institutional conformism required by totalising institutions such as the armed forces.'[76] Some of these trends

70 Caddick-Adams, 'The Volunteers', 95; Strachan, 'The Civil-Military Gap in Britain', 43–63.
71 Edmunds et al., 'Reserve forces and the transformation of British military organisation', 126.
72 YouGov (2015) *Report on British attitudes to defence, security and the armed forces*, available at https://yougov.co.uk/news/2014/10/25/report-british-attitudes-defence-security-and-arme/, retrieved 9 September 2016.
73 Houghton, N. (2014) 'Annual Chief of the Defence Staff Lecture' at RUSI London.
74 Edmunds et al., 'Reserve forces', 131.
75 Ibid., 127.
76 Ibid.

are worthy of further examination here. For example, in the 1950s the Labour and Conservative parties had a combined total of over four million members, today Labour have an estimated 540,000 while the Conservatives 124,000.[77] The average age of Conservative Party membership is 68.[78] Similarly, only 13 per cent of people report going to a religious service once a week or more and the Church of England's own attendance figures also attest to decline; in 2013 average Sunday attendance figures were just 785,000, half the number that attended in 1968.[79] Interestingly, this precisely fits the decline of social capital identified by Putnam from the mid-1960s onwards, which has been mirrored by declining participation in the TA and Army Reserve. While these are nonetheless traditional methods of assessing social capital, other trends are emerging. Today the National Trust has four million members and Sky TV ten million subscribers.[80] Social media use amongst the Y and Z Generation is regarded as contributing to a ten-fold increase in Narcissistic Personality Disorder, while there has been a recent decline in sports participation and gym use in Britain, indicating that heightened individualism has potential pitfalls and as a society we are getting less fit.[81] Complementing Edmunds et al., and supporting Putnam's and Inglehart's findings, it seems clear that a profound change in the nature of British society has occurred. The implications of these changing societal values – and indeed changing British demographics – for army and reserve recruitment have been identified by Carter, who made the vision of an 'inclusive' army that is more sensitive to equality and diversity, and more flexible in terms of employment models, major themes of his tenure as CGS.[82] But while Edmunds et al. highlight the changed nature of British society, they do not seek to explain the sources of this change.

Numerous British authors have charted how the post-war political consensus which defined the relationship between the British state and its people from 1945–79 has been gradually undone by the neo-liberal political ideology of successive Conservative and Labour governments since that date. David Marquand has referred to the post-war consensus as lasting 'from the mid-1940s to the mid-1970s, [when] most of [Britain's] political class shared a tacit governing philosophy which might be called "Keynesian social-democracy"'.[83] Both Labour and the Conservatives 'generally accepted [the] values and assumptions' of a 'three-fold commitment

77 House of Commons (2018) 'Membership of UK Political Parties', Briefing Paper, 1 May.

78 Jones, O. (2015) *The Establishment*, London: Penguin, 69.

79 Figures available at https://www.churchofengland.org/about-us/facts-stats/research-statistics. aspx, retrieved 25 May 2016.

80 Lt Gen. Sir Paul Newton, Lecture, Exeter Security and Strategy Institute, 6 June 2016.

81 *The Guardian*, 17 March 2016, 'I, narcissist – vanity, social media, and the human condition'; *The Daily Telegraph*, 11 June 2015 'Peak physique? Britain's gym bubble bursts'.

82 Ministry of Defence (2015) 'Vision for an Inclusive Army' available at https://modmedia.blog. gov.uk/2015/08/03/army-chief-sets-out-vision-for-inclusive-future/, retrieved 12 September 2016.

83 Marquand, D. (1988) *The Unprincipled Society*, London: Fontana, 2.

to full employment, to the welfare state, and to the co-existence of large public and private sectors in the economy'.[84] This political consensus had been primarily generated by the sacrifices of the British 'Great Generation' in the Second World War, resulting in Clement Attlee's famous victory over Winston Churchill in the 1945 election.[85] Riding a tide of popular support, Attlee's government followed Keynesian economic policies aimed at high rates of employment, nationalising public utilities and major industries, and greatly enlarging the system of social services, including establishing the NHS. Andrew Gamble has argued that the success of the wartime coalition government was also an important factor in generating this consensus, while Peter Clarke has detailed how this extension of free health care to all citizens also had a moral component, increasing social equality between the classes.[86] Trade unions also remained strong and a major influence on politics. In Britain, this post-war consensus involving greater state intervention in the economy and the greater provision of social services was accepted by both major political parties for over three decades.

However, in the late 1960s, this consensus began to be undermined by Britain's increasingly poor economic performance, evidenced in the decreasing competitiveness of British industry, low growth rates, and, especially after the 1973 oil shock, increasing inflation and unemployment. Cumulatively and gradually, these caused the development of an institutional crisis as successive governments' interventions in the economy failed. Marquand argues that these failures began a process of 'ideological polarisation which destroyed the post-war consensus.'[87] In his study of the rise of 'New-Right', Gamble shows how the Thatcherites that embodied this reinvigorated liberalism began to argue that the Keyneisan 'social-democratic polices had led to the morass of inflation, mass unemployment, excessive taxation and a swollen public sector.'[88] Crucially, Gamble argues that:

> The particular quarrel of the Thatcherites was with the attitudes and policies to which conservatives had become committed in the 1940s and 1950s. It was their acquiescence in the social democratic hegemony that they wished to change.[89]

These goals started to be realised when Margaret Thatcher was elected prime minister in 1979 and began pursuing economic policies primarily aimed at reducing

---

84 Ibid., 3.

85 Ibid., 21.

86 Gamble, A. (1994) *The Free Economy and the Strong State: The Politics of Thatcherism*, 2nd Edition, London: Macmillan, 70; Clarke, P. (1978) *Liberals and Social Democrats*, Cambridge: Cambridge University Press.

87 Marquand, *The Unprincipled Society*, 47.

88 Gamble, *The Free Economy and the Strong State*, 35.

89 Ibid., 69.

inflation and de-regulating the markets rather than maintaining high employment. Numerous policies concerning taxation, local authority reform, and the sale of nationalised industries incrementally but determinedly undid the post-war consensus.[90] These were couched in arguments about rising living standards and efficient economy. These policies led to major social changes in the UK, including the breaking of the powerful trade unions and the privatisation of industry, but also coincided with rising living standards and more rapid, but less stable, economic growth. Meanwhile, the end of the Fordist mode of production and growth of information technology diversified and atomised work forces. Similarly, the share of income going to the top 10 per cent of the UK population rose from 21 per cent in 1979 to 31 per cent in 2009, reversing a deeper negative trend during 1938–79 (closely mirroring the post-war consensus period) and indicating a concentration of wealth that potentially undermines social fabric.[91] These outcomes were also linked with rising materialism in the UK – most lavishly embodied by the rise of the city after Thatcher's 'Big Bang' de-regulations of 1986 – but also the Conservatives' normative argument about the importance of individual freedom and motivation in society. Indeed, Thatcher's view on the matter is worthy of quoting here:

> They are casting their problems at society. And, you know, there's no such thing as society. There are individual men and women and there are families. And no government can do anything except through people, and people must look after themselves first. It is our duty to look after ourselves and then, also, to look after our neighbours.[92]

Crucially, after the Conservatives lost power in 1997, privatisation continued under Blair's – and indeed sociologist Anthony Giddens' – 'Third Way', reliant as it was on the narrative that due to globalisation state intervention in market forces was fallacy.[93] While these approaches generally increased employment and living standards, under Cameron and May, the reformation of the state's role in British society deepened and quickened. There have been major reductions in state spending in almost every department and attempts to further privatise the education and health sectors around Adam Smith's principle of the 'invisible hand' in the drive for efficiencies. This time, the austerity narrative was utilised to justify the use of this 250-year-old guiding principle and mask the political ideology behind it, in spite of the fact that the de-regulation of the banks was a major contributing

---

90 Ibid., 123, 126.

91 The Equality Trust (2014) 'How Has Inequality Changed?' available at https://www. equalitytrust.org.uk/how-has-inequality-changed, retrieved 16 July 2016.

92 *The Guardian*, 8 April 2013, 'Margaret Thatcher: A life in quotes'.

93 Watson, M. and Hay, C. (2003) 'The discourse of globalisation and the logic of no alternative: rendering the contingent necessary in the political economy of New Labour', *Policy and Politics*, 31(3).

factor to the 2008 global recession. Indeed, after the cuts to Britain's armed forces in 2010–11, and the threat of major cuts in 2018, many questioned whether the state now retained the capability to protect its citizens, indicating that the Hobbesian contract between it and the population may be under threat.[94] Thus, the decline in social capital and the rise of individualism in British society has been accompanied by a gradual, but cumulatively profound, parallel change in the nature of the relationship between British society and the state.

It is perhaps not too much to suggest that the neo-liberal values of successive British governments have contributed to the dismantling of post-war consensus underpinned by the Great Generation's sacrifices, the profound re-organisation of the British state in the last three decades, and the rise of individualism. This, in turn, appears to have curtailed the new Army Reserves' ability to recruit to (a historically minute) establishment today. Of course, it is arguable that the nature of the relationship between the British military and citizens also changed with the phased ending of conscription and national service in 1960 and the move to professionalism. Indeed, the army has also changed, and the recent standardisation of Army Reserve fitness and medical requirements with those of the regulars has definitely impacted recruitment.[95] However, based on recruitment and retention rates, if FR20 had been attempted in 1965 – by which time conscription and national service had ended and a fully professional army established – the figures cited above suggest that the Army Reserve could have filled 11,000 vacancies in six months to a year, rather than struggling to do so in six years. The Conservatives themselves recognised that the relationship between the state and citizens had changed in FR20: 'We all depend on national security; however, most people choose not to contribute to it beyond paying their taxes.'[96] As a result of the pursuit of neo-liberalist policies, paying taxes is increasingly viewed as the sole civic duty of citizens. Thus, the Conservative government's call for a greater volunteer ethos representative of the commitment of Great Generation's citizen-soldier appears particularly ironic as not only has society changed fundamentally since then, but it has done so as a result of neo-liberal ideals still championed by the very government that made the call. The Conservative Party instigated and oversaw both the deep reduction of state involvement in society, and contributed to a change in social attitudes, yet paradoxically expects citizens to flock to the Army Reserve in order to compensate for their defence cuts. It is perhaps significant that this trend has only been reversed with the introduction of major monetary incentives, which have in turn raised concerns at both the senior and junior levels about the quality and commitment of personnel recruited under these terms.

---

94 *BBC News*, 17 September 2016, 'UK military ill-prepared to defend an attack, says retired chief'.

95 Francois, M. (2017) 'Filling the Ranks': A Report for the Prime Minister on the State of Recruiting into the United Kingdom Armed Forces, 6.

96 *Future Reserves 2020*, 7.

In the final analysis, while an undoubtedly well-meaning attempt to end the neglect of and revitalise the TA, FR20 was too ambitious in its attempt to routinely and regularly deploy the new Army Reserve at the unit or sub-unit level as part of the army's wider operational readiness cycle. The difficulties that the FR20 transformation experienced highlight that the political motivations for transformation must be supported by senior *and* mid-level commanders if they are to be successful. They also must not ignore the distinctive institutional character of the organisations they seek to transform. In short, they must be fully workable. Politics not only drives top-down transformations, it can also limit them by ignoring the reality at the bottom of the organisations they seek to transform. If this happens, supporting Allison's arguments, what I label a 'partial transformation' may occur, where some aspects are successful but others that are not are quietly jettisoned. Meanwhile, I have detailed that both logistics and reserve forces transform in broadly similar ways to that identified in the extant literature. Overall, by transforming through top-down, bottom-up and emulative practices spanning both structural and normative/cultural divides, the evidence supports Foley's et al.'s arguments on complementary sources of military change. Nevertheless, the evidence presented here suggests that transformations of reserve forces are likely to take longer to succeed than in regular forces due to their distinctive part-time nature, their potential threat to the army's organisational survival, and their closer proximity to society that brings with it political advantages. For this reason, transforming reserve organisations can prove more difficult to reform than regular forces. However, there is one final, cautionary observation to be made. As the evidence above suggests, for peacetime military transformations to succeed in Western democratic states, they must also be grounded in the realities of modern society. Failing to understand how society has been changed by the result of policies that themselves sought to transform society, is failing to identify the nature of society that can make military transformations successful or not, especially in the case of reserve forces. It seems obvious, but ultimately, and even in regular professional forces, in an increasingly individualistic era, military transformations must carefully consider wider societal transformation if they are to be fully successful.

Bibliography

Allison, G. (1971) *The Essence of Decision: Explaining the Cuban Missile Crisis*, Boston: Little and Brown.

Argyris, A. and Schön, D. (1995) *Organizational Learning II: Theory, Method, and Practice*, Reading, MA: Addison-Wesley.

Avant, D. (1994) *Political Institutions and Military Change: Lessons from Peripheral Wars*, New York: Cornell University Press.

—— (2005) *The Market for Force*, Cambridge: Cambridge University Press.

Bailey, J., Iron, R. and Strachan, H. (eds) (2013) *British Generals in Blair's Wars*, London: Routledge.

Bancroft, A. (2009) *Drugs, Intoxication and Society*, Cambridge: Polity.

Barnett, C. (1980) 'Radical Reform 1902–14', in Perlmutter, A. and Bennett, V. (eds) *The Political Influence of the Military: A Comparative Reader*, New Haven: Yale University Press.

Bartone, P. and Kirkland, F. (1991) 'Optimal Leadership in Small Army Units', in Gal, R. and Mangelsdorff, A. (eds) *Handbook of Military Psychology*, Chichester: Wiley.

Baudrillard, J. (1994) *Simulacra and Simulation*, Michigan: University of Michigan Press.

Bauman, Z. (2006) *Liquid Times: Living in an Age of Uncertainty*, Cambridge: Polity.

Beal, D., Cohen, R., Burke, M. and McLendon, C. (2003) 'Cohesion and Performance in Groups: A Meta-Analytic Clarification of Construct Relations', *Journal of Applied Psychology*, 88(6).

Beckett, I. (1982) *Riflemen Form: A Study of the Rifleman Volunteer Movement, 1859–1908*, Aldershot: The Ogilby Trust.

Beckett, I. (2008) *Territorials: A Century of Service*, Plymouth: DRA Publishing.

Ben-Dor, G., Pedazhur, A., Canetti-Nisim, D., Zaidise, E., Perliger, A. and Bermanis, S. (2008) '"I versus We" Collective and Individual Factors of Reserve Service Motivation during War and Peace', *Armed Forces and Society*, 34(4).

Ben-Ari, E. and Lomksy-Feder, E. (2011) 'Epilogue: Theoretical and Comparative Notes on Reserve Forces', *Armed Forces and Society*, 37(2).

Ben-Shalom, U., Lehrer, Z. and Ben-Ari, E. (2005) 'Cohesion During Military Operations: A field study on combat units in the Al-Aqsa Intifada', *Armed Forces and Society*, 32(1).

Berdal, M. (2003) 'How "New" are "New Wars"? Global Economic Change and the Study of Civil War', *Global Governance*, 9(4).

Berkshire Consultancy (2010) *Study of Women in Combat – Investigation of Quantitative Data.*

Biddle, S. (2006) *Military Power: Explaining Victory and Defeat in Modern Battle*, Princeton: Princeton University Press.

Biddulph, R. (1904) *Lord Cardwell at the War Office: A History of His Administration 1868–1874*, London: John Murray.

Bondy, W. (2001) 'Postmodernism and the Source of Military Strength in the Anglo-West', *Armed Forces and Society*, 31(1).

Brazier, J. (2012) '"All Sir Garnet!" Lord Wolseley and The British Army in the First World War', *Military History Monthly*, May.

British Army (2013) *Battlefield Equipment Support Doctrine.*

—— (2014) *Tactical Logistics Support Handbook.*

—— (2016) *Soldier Magazine*, August.

Burk, J. (2007) 'The Changing Moral Contract for Military Service', in Bacevich, A. (ed.) *The Long War: America's Quest for Security since World War II*, New York: Columbia University Press.

Bury, P. (2010) *Callsign Hades*, London: Simon & Schuster.

—— (2015, unpublished) *Report on Recruitment and Retention*, prepared for CD CSS, 12 January 2015.

—— (2017) 'Recruitment and Retention in British Army Logistics Units', *Armed Forces and Society*, 43(4).

—— (2016) 'Barossa Night: Cohesion in the British Army Officer Corps', *The British Journal of Sociology*, 68(2).

—— (2017) 'The Changing Nature of Reserve Cohesion: A Study of Future Reserves 2020 and British Army Reserve Logistic Units', *Armed Forces and Society*, available at https://doi.org/10.1177/0095327X17728917.

—— (2018) 'Future Reserves 2020: Perceptions of Cohesion, Readiness and Transformation in the British Army Reserve', *Defence Studies*, 18(4).

Bury, P. and King, A. (2015) 'A Profession of Love? Cohesion in a British Platoon in Afghanistan', in King, A. (ed.) *Frontline: Combat and Cohesion in the Twenty-First Century*, Oxford: Oxford University Press.

Caddick-Adams, P. (2002) 'The Volunteers', in Alexandrou, A., Bartle, R. and Holmes, R. (eds) *New People Strategies for the British Armed Forces*, London: Frank Cass.

Carter, General N. (2014) 'Army 2020 The Army Reserve', *Army Reserve Quarterly*, Autumn.

Catignani, S. (2012) 'Getting COIN at the Tactical Level in Afghanistan: Re-Assessing Counter-Insurgency Adaptation in the British Army', *Journal of Strategic Studies*, 35(4).

Catignani, S. (2014) 'Coping with Knowledge: Organisational Learning in the British Army?', *Journal of Strategic Studies*, 37(1).

Cawkwell, T. (2015) *UK Communication Strategies for Afghanistan, 2001–2014*, Farnham: Ashgate Publishing.

Christopher, M. (1998) *Logistics and Supply Chain Management*, London: Pitman.

Christopher, M. and Holweg, M. (2011) '"Supply Chain 2:0": managing supply chains in the era of turbulence', *International Journal of Physical Distribution and Logistics Management*, 41(1).

Cilliers, P. (2000) 'Rules and Complex Systems', *Emergence*, 40(3).

Clarke, P. (1978) *Liberals and Social Democrats*, Cambridge: Cambridge University Press.

Clausewitz, C. von (1989) *On War* (ed. and trans. Howard, M. and Paret, P.), Princeton: Princeton University Press.

Cohen, E. (2001) 'Twilight of the Citizen-Soldier', *Parameters*, Summer.

—— (2004) 'Change and Transformation in Military Affairs', *Journal of Strategic Studies*, 27(3).

Connelly, V. (2013) *Cultural Differences between the Regular Army and TA as Barriers to Integration*, Unpublished paper prepared for Director, Personnel, MoD.

Cooley, C. (1909) *Social Organization: A Study of the Larger Mind*, New York: Charles Scribner's Sons.

Cornish, P. and Dorman, A. (2011) 'Dr Fox and the Philosopher's Stone: the alchemy of national defence in the age of austerity', *International Affairs*, 87(2).

—— (2013) 'Fifty shades of purple? A risk-sharing approach to the 2015 Strategic Defence and Security Review', *International Affairs*, 89(5).

—— (2014) *Reserve Forces External Scrutiny Team Annual Report*, London: NAO.

—— (2015) *Reserve Forces External Scrutiny Team Annual Report*, London: NAO.

—— (2016) *Future Reserves 2020: 2016 External Scrutiny Team Report*.

—— (2017) *Future Reserves 2020: 2017 External Scrutiny Team Report*.

—— (2018) *Future Reserves 2020: 2018 External Scrutiny Team Report*.

Cunningham, H. (1975) *The Volunteer Force: A Social and Political History, 1859–1908*, London: Croom Helm.

Cusumano, E. (2016) 'Bridging the Gap: Mobilisation Constraints and Contractor Support to US and UK Military Operations', *Journal of Strategic Studies*, 39(1).

Dandeker, C., Booth, B., Kestnbaum, M. and Segal, D. (2001) 'Are Post-Cold War Militaries Postmodern?', *Armed Forces and Society*, 27(3).

Dandeker, C., Greenberg, N. and Orme, G. (2011) 'The UK's Reserve Forces: Retrospect and Prospect', *Armed Forces and Society*, 37(2).

Dannatt, General Sir R. (2011) *Leading from the Front*, London: Penguin.

Davids, C., Beeres, R. and van Fenema, P. (2013) 'Operational defense sourcing: organizing military logistics in Afghanistan', *International Journal of Physical Distribution & Logistics Management*, 43(2).

Defence Select Committee (2014) *Ninth Report: Future Army 2020*, London: TSO.

Demchak, C. (2002) 'Complexity and Theory of Networked Militaries', in Farrell, T. and Terriff, T. (eds) *The Sources of Military Change: Culture, Politics Technology*, Boulder, CO: Lynne Rienner.

——— (2003) 'Creating the Enemy: Global Diffusion of the Information Technology-Based Military Model', in Goldman, E. and Eliason, L. (eds) *The Diffusion of Military Technology and Ideas*, Stanford, CA: Stanford University Press.

Dennis, P. (1987) *The Territorial Army 1907–1940*, Suffolk: Royal Historical Society.

Dibella, A., Nevis, E. and Gould, J. (1996) 'Understanding Organizational Learning Capability', *Journal of Management Studies*, 33(3).

Dimaggio, P. and Powell, W. (1983) 'The Iron Cage Revisited: Institutional isomorphism and collective rationality in organisational fields', *American Sociological Review*, 48(2).

Dion, K. (2000) 'Group Cohesion from "Field of Forces" to Multi-Dimensional Construct', *Group Dynamics: Theory, Research and Practice*, 4(1).

Duffield, M. (2001) *Global Governance and the New Wars*, London: Zed Books.

Duncan, S. (1997) *Citizen Warriors: America's National Guard and Reserve Forces and the Politics of National Security*, Novato, CA: Presidio Press.

Edmunds, T. (2010) 'The Defence Dilemma in Britain', *International Affairs*, 86(2).

——— (2012) 'British civil–military relations and the problem of risk', *International Affairs*, 88(2).

Edmunds, T., Dawes, A., Higate, P., Jenkins, N. and Woodward, R. (2016) 'Reserve forces and the transformation of British military organisation: soldiers, citizens and society', *Defence Studies*, 16(2).

Egnell, R. (2006) 'Explaining US and British Performance in Complex Expeditionary Operations: The Civil-Military Dimension', *The Journal of Strategic Studies* 29(6).

——— (2011) 'Lessons from Helmand, Afghanistan: What now for British Counterinsurgency', *International Affairs*, 87(2).

Engels, D. (1978) *Alexander the Great and the Logistics of the Macedonian Army*, Berkeley: University of California Press.

Erbel, M. and Kinsey, C. (2015) 'Think Again – Supplying War: Re-appraising Military Logistics and Its Centrality to Strategy and War', *Journal of Strategic Studies*, 41(4).

Evans-Pritchard, E. (1940) *The Nuer: A Description of the Modes, Livelihood and Political Institutions of a Nilotic People*, Oxford: Clarendon Press.

Farrell, T. (2010) 'Improving in War: Military Adaptation and the British in Helmand Province, Afghanistan, 2006–2009', *Journal of Strategic Studies*, 33(4).

Farrell, T. and Terriff, T. (eds) (2001) *The Sources of Military Change: Culture, Politics, Technology*, Boulder, CO: Lynne Rienner.

Farrell, T., Rynning, S. and Terriff, T. (2013) *Transforming Military Power since the Cold War: Britain, France, and the United States, 1991–2012*, Cambridge: Cambridge University Press.

Feaver, P. (2003) *Armed Servants: Agency, Oversight, and Civil–Military Relations*, Cambridge, MA: Harvard University Press.

Festinger, L., Back, K. and Schacter, S. (1950) *Social Pressures in Informal Groups: A Study of Human Factors in Housing*, New York: Harper.

Fletcher, M. (2011) 'Joint Supply Chain Architecture', *Army Sustainment*, 43(3).

Foley, R., Griffin, S. and McCartney, H. (2011) '"Transformation in Contact": learning the lessons of modern war', *International Affairs*, 87(2).

Francois, M. (2017) *'Filling the Ranks': A Report for the Prime Minister on the State of Recruiting into the United Kingdom Armed Forces*.

Fraser, P. (1973) *Lord Esher: A Political Biography*, London: Hart-Davis.

French, D. (2005) *Military Identities: The Regimental System, the British Army and the British People*, Oxford: Oxford University Press.

Gal, R. (1986) 'Unit Morale: From a Theoretical Puzzle to an Empirical Illustration – An Israeli Example', *Journal of Applied Social Psychology*, 16(6).

Gamble, A. (1994) *The Free Economy and the Strong State: The Politics of Thatcherism*, 2nd Edition, London: Macmillan.

Geertz, C. (1984) 'From the Native's Point of View: on the nature of anthropological understanding', in Shweder, R. and LeVine, R. (eds) *Culture Theory: Essays on Mind, Self, and Emotion*, New York: Cambridge University Press.

Giddens, A. (1990) *The Consequences of Modernity*, Cambridge: Polity.

Glas, A., Hofmann, E. and Essig, M. (2013) 'Performance-based logistics: a portfolio for contracting military supply', *International Journal of Physical Distribution & Logistics Management*, 43(2).

Griffin, S. (2016) 'Military Innovation Studies: Multidisciplinary or Lacking Discipline?', *Journal of Strategic Studies*, 40(1–2).

Griffith, J. (1988) 'Measurement of Group Cohesion in U.S. Army Units', *Basic and Applied Social Psychology*, 9(2).

—— (2002) 'Multilevel Analysis of Cohesion's Relation to Stress, Well-Being, Identification, Disintegration and Perceived Combat Readiness', *Military Psychology*, 14(3).

—— (2007) 'Further Considerations Concerning the Cohesion-Performance Relation in Military Settings', *Armed Forces and Society*, 34(1).

—— (2007) 'Institutional Motives for Serving in the U.S. Army National Guard', *Armed Forces and Society*, 34(2).

—— (2009) 'After 9/11 What Kind of Reserve Soldier', *Armed Forces and Society*, 35(2).

—— (2009) 'Being a Reserve Soldier: A Matter of Social Identity', *Armed Forces and Society*, 36(1).

—— (2011) 'Contradictory and Complementary Identities of US Army Reservists: A Historical Perspective', *Armed Forces and Society*, 37(2).

_____ (2011) 'Decades of transition for the US reserves: Changing demands on reserve identity and mental well-being', *International Review of Psychiatry*, 23(2).

_____ (2012) 'Correlates of Suicide Among Army National Guard Soldiers', *Military Psychology*, 24.

Grissom, A. (2006) 'The Future of Military Innovation Studies', *Journal of Strategic Studies*, 29(5).

_____ (2013) 'Shoulder-to-Shoulder Fighting Different Wars: NATO Advisors and Military Adaptation in Afghan National Army, 2001–2011', in Farrell, T., Osinga, F. and Russel, J. (eds) *Military Adaptation in Afghanistan*, Stanford: Stanford University Press.

Haldane, R. (1929) *Richard Burdon Haldane. An Autobiography*, 2nd edition, London: Hodder and Stoughton.

Harkness, K. and Hunzeker, M. (2015) 'Military Maladaptation: Counter-insurgency and the Politics of Failure', *Journal of Strategic Studies*, 38(6).

Held, D., McGrew, A., Goldblatt, D. and Perraton, J. (1999) *Global Transformations*, Cambridge: Polity.

Henderson, D. (1985) *Cohesion: The Human Element*, Washington, DC: National Defence University Press.

Herodotus (2008) *The Histories* (trans. R. Wakefield), Oxford: Oxford Paperbacks.

Higgens, S. (2010) 'How was Richard Haldane able to reform the British Army?', Unpublished MPhil. dissertation, University of Birmingham.

HM Government (1965) *Sir Philip Allen's Home Defence Review Committee Report*, London: HMSO.

HM Government (2010) *Securing Britain in an Age of Uncertainty: The Strategic Defence and Security Review*, London: HMSO.

_____ (2015) *National Security Strategy and Strategic Defence and Security Review 2015: A Secure and Prosperous United Kingdom*, Norwich: HMSO.

Hogg, M. (1992) *The Social Psychology of Group Cohesiveness: From Attraction to Social Identity*, New York: Harvester Wheatsheaf.

Horowitz, D. and Kimmerling, B. (1974) 'Some Social Implications of Military Service and the Reserves System in Israel', *Archives Européennes de Sociologie*, 15(2).

House of Commons (2012) Library Report 'Army 2020'.

_____ (2015) 'Membership of UK Political Parties', Briefing Paper, 11 August.

House of Commons Defence Committee (2010) *First Report, The Strategic Defence and Security Review*, London: TSO.

_____ (2000) *Second Report: Ministry of Defence Annual Reporting Cycle*, London: HMSO.

Howard, M. (1967) *Lord Haldane and the Territorial Army*, London: Birkbeck College.

Hughes, G. (2013) 'British Generals in Blair's Wars: A Review Article', *The Round Table*, 102(6).

Huntington, S. (1957) *The Soldier and the State: The Theory and Politics of Civil–Military Relations*, Cambridge, MA: Harvard University Press.

Husserl, E. (1970) *The Crisis of European Sciences and Transcendental Philosophy* (trans. D. Carr) Evanston, IL: Northwestern University Press.

Inglehart, R. (1977) *The Silent Revolution*, Princeton: Princeton University Press.

—— (1990) *Culture Shift in Advanced Industrial Society*, Princeton: Princeton University Press.

Ingraham, L. and Manning, F. (1981) 'Cohesion: Who Needs It, What Is It, and How Do We Get It to Them?', *Military Review*, 61(6).

Janowitz, M. (1971) *The Professional Soldier: A Social and Political Portrait*, New York: Free Press.

Jones, O. (2015) *The Establishment*, London: Penguin.

Kaldor, M. (1999) *New and Old Wars: Organised Violence in the Global Era*, Cambridge: Polity.

Kaldor, M., Albrecht, U. and Schméder, G. (eds), (1998) *The End of Military Fordism*, London: Pinter.

Kane, T. (2001) *Military Logistics and Strategic Performance*, London: Routledge.

Kaufmann, H. (1971) *The Limits of Organisational Change*, Tuscaloosa: University of Alabama Press.

Kellner, P. (2012) 'Public Perceptions of the Army', RUSI Land Warfare Conference, 3 July 2013, available at https://www.youtube.com/watch?v=eS6DgD5ZUSM.

Kent, J. and Flint, D. (1997) 'Perspectives on the Evolution of Logistics Thought', *Journal of Business Logistics*, 18(2).

Kier, E. (1997) *Imagining War: French and British Military Doctrine*, Princeton: Princeton University Press.

King, A. (2005) 'Towards a Transnational Europe: The Case of the Armed Forces', *European Journal of Social Theory*, 8(4).

—— (2006a) 'The Post-Fordist Military', *Journal of Political and Military Sociology*, 34(2).

—— (2006b) 'The Word of Command: communication and cohesion in the military', *Armed Forces and Society*, 32(1).

—— (2009) 'The Special Air Service and the Concentration of Military Power', *Armed Forces and Society*, 35(4).

—— (2010) 'Understanding the Helmand campaign: British military operations in Afghanistan', *International Affairs*, 86(2).

—— (2011) *The Transformation of Europe's Armed Forces: From the Rhine to Afghanistan*, Cambridge: Cambridge University Press.

—— (2013) *The Combat Soldier: Infantry Tactics and Cohesion in the Twentieth and Twenty-First Centuries*, Oxford: Oxford University Press.

—— (2015) 'Discipline and Punish: Encouraging Combat Performance in the Citizen and Professional Army', in King, A. (ed.) *Frontline: Combat and Cohesion in the Twenty-First Century*, Oxford: Oxford University Press.

_____ (2019) *Command: The Twenty-First Century General*, Cambridge: Cambridge University Press (forthcoming).

Kinsey, C. (2005) 'Regulation and Control of Private Military Companies', *Contemporary Security Policy*, 26(1).

Kinzer Stewart, N. (1991) *Mates and Muchacos: Unit Cohesion in the Falklands/ Malvinas War*, New York: Brassey's.

Kirke, C. (2008) 'Issues in integrating Territorial Army Soldiers into Regular British Units for Operations: A Regular View', *Defense and Security Analysis*, 24(2).

Koss, S. (1969) *Lord Haldane: Scapegoat for Liberalism*, New York: Columbia University Press.

Krahmann, E. (2005) 'Security Governance and the Private Military Industry in Europe and North America', *Conflict, Security and Development*, 5(2).

Krebs, R. (2009) 'The Citizen-Soldier Tradition in the United States: Has its Demise Been Greatly Exaggerated?', *Armed Forces and Society*, 36(1).

Levinson, M. (2006) *The Box*, Princeton: Princeton University Press.

Levitt, B. and March, J. (2007) 'Organizational Learning', *Annual Review of Sociology*, 14.

Levy, Y. (2010) 'The Essence of the "Market Army"', *Public Administration Review*, 70(3).

Libicki, M. (1996) 'The Emerging Primacy of Information', *Orbis*, 40(2).

Lindsay, C. (2003) 'A Century of Labour Market Change', Report for Office of National Statistics.

Lomsky-Feder, E., Gazit, N. and Ben-Ari, E. (2008) 'Reserve Soldiers as Transmigrants: Moving between the Civilian and Military Worlds', *Armed Forces and Society*, 34(4).

Louth, J. (2015) 'Logistics as a Force Enabler: The Future Operational Imperative', *RUSI Journal*, 160(3).

Lynn, J. (1993) *Feeding Mars: Logistics in Western Warfare from the Middle Ages to the Present*, Boulder, CO: Westview Press.

Lyotard, J-F. (1979) *The Postmodern Condition: A Report on Knowledge*, Manchester: Manchester University Press.

MacCoun, R. (1993) 'What is known about unit cohesion and military performance', in *Sexual Orientation and U.S. Military Personnel Policy: Options and Assessment*, Washington: RAND.

MacCoun, R., Kier, E. and Belkin, A. (2006) 'Does Social Cohesion Determine Motivation in Combat?', *Armed Forces and Society*, 32(4).

MacCoun, R. and Hix, W. (2010) 'Cohesion and performance', in National Defense Institute, *Sexual Orientation and U.S. Military Personnel Policy: An Update of RAND's 1993 Study*. Santa Monica: RAND Corporation.

Maddox, E. (2005) 'Organizing Defense Logistics: What Strategic Structures Should Exist for the Defense Supply Chain', unpublished thesis.

Malinowski, B. (1929) *The Sexual Life of Savages in North-Western Melanesia: An Ethnographic Account of Courtship, Marriage and Family Life Among the*

Natives of the Trobriand Islands, British New Guinea, New York: Halcyon House.

Mallinson, A. (2011) *The Making of the British Army*, London: Bantam.

Marcus, R. (2015) 'Military Innovation and Tactical Adaptation in the Israel-Hizballah Conflict: The Institutionalization of Lesson-Learning in the IDF', *Journal of Strategic Studies*, 38(4).

Marlowe, D. (1985) 'New Manning System field evaluation' (Tech. Rep. No. 1). Washington, DC: Walter Reed Army Institute of Research, Department of Military Psychiatry.

Marquand, D. (1988) *The Unprincipled Society*, London: Fontana.

Mentzer, J., Soonhong, M. and Bobbit, M. (2004) 'Toward a Unified Theory of Logistics', *International Journal of Physical Distribution and Logistics Management*, 34(8).

Mervin, K. (2005) *Weekend Warrior: A Territorial Soldier's War in Iraq*, Edinburgh: Mainstream.

Ministry of Defence (2004) *Operation TELIC – United Kingdom Military Operations in Iraq*, London: HMSO.

—— (2011) *Future Reserves 2020 – The Independent Commission to Review the United Kingdom's Reserve Forces*, London: HMSO.

—— (2012) *Future Reserves 2020: Delivering the Nation's Security Together* (Green Paper) London: HMSO.

—— (2012) *Joint Service Publication 886, Vol. 1 The Defence Logistics Support Chain*, London: MoD.

—— (2012) *Transforming the British Army, July 2012 – Modernising to Face an Unpredictable Future*, London: MoD.

—— (2013) *Defence Estate Rationalisation Update.*

—— (2013) *Reserves in the Future Force 2020: Valuable and Valued* (White Paper) London: HMSO.

—— (2013) *Transforming the British Army, An Update – July 2013*, London: MoD.

—— (2014) *Joint Doctrine Publication 0-01* (JDP 0-01), 5th Edition, Swindon: Development, Concepts and Doctrine Centre.

—— (2015) *Joint Doctrine Publication 4.0 Logistics for Joint Operations*, 4th Edition, Swindon: Development, Concepts and Doctrine Centre.

—— (2015) *Reserve Continuous Attitudes Survey 2015.*

—— (2016) *Armed Forces Continuous Attitudes Survey 2016.*

—— (2016) *Reserve Continuous Attitudes Survey 2016.*

—— (2018) *Reserve Continuous Attitudes Survey 2018.*

Mitchinson, K. (2005) *Defending Albion: Britain's Home Army 1908–1919*, London: Palgrave.

—— (2008) *England's Last Hope: The Territorial Force 1908–14*, London: Palgrave.

—— (2014) *The Territorial Force at War 1914–16*, London: Palgrave.

Mooney, J. and Crackett, J. (2018) 'A Certain Reserve: Strategic Thinking and Britain's Army Reserve', *The RUSI Journal*, 163(4).

Moore, D. and Antill, P. (2000) 'Where Do We Go From Here? Past, Present and Future Logistics of the British Army', *The British Army Review*, 125.

Morgan, D. (1988) *Focus Groups as Qualitative Research*, London: Sage.

Morris, A. (1971) 'Haldane's Army Reforms 1906–8: The Deception of the Radicals', *History*, 56(186).

Moskos, C. (1977) 'From Institution to Occupation, Trends in Military Service', *Armed Forces and Society*, 4(1).

Moskos, C. and Burk, J. (1994) 'The Post-Modern Military', in Burk, J. (ed.), *The Military in New Times: Adapting Armed Forces to a Turbulent World*, Boulder, CO: Westview Press.

Moskos, C. and Wood, F. (eds) (1988) 'Introduction', in *The Military: More than Just a Job?* New York: Pergamon-Brassey.

Moskos, C., Williams, J. and Segal, D. (eds) (2000) *The Postmodern Military: Armed Forces after the Cold War*, Oxford: Oxford University Press.

Mullen, B. and Copper, C. (1994) 'The Relation between Group Cohesiveness and Performance: An Integration', *Psychological Bulletin*, 115(2).

Nadler, D. and Tushman, M. (1997) *Competing by Design*, New York: Oxford University Press.

National Audit Office (2005) *Assessing and Reporting Military Readiness*.

—— (2014) *Report on Army 2020: Executive Summary*, 5, London: HMSO.

NATO *Logistics Handbook 2012*.

O'Hanlon, M. (2009) *The Science of War*, Princeton: Princeton University Press.

O'Konski, M. (1999) 'Revolution in Military Logistics: An Overview', *Army Logistician*, Jan.–Feb. 1999.

Office of National Statistics (2013) 'Labour Market Statistics, May 2013'.

Oliver, L., Harman, J., Hoover, E., Hayes, S. and Pandhi, N. (1999) 'A Qualitative Integration of the Military Cohesion Literature', *Military Psychology*, 11(1).

Olson, R. and Scrogin, T. (1974) 'Containerisation and Military Logistics', *The Journal of Maritime Law and Commerce*, 6(1).

Oxford English Dictionary (1983) Oxford: Oxford University Press.

Oxford English Dictionary (2008) Oxford: Oxford University Press.

Pagonis, W. (1992) *Moving Mountains*, Boston: Harvard Business School Press.

Perliger, A. (2011) 'The Changing Nature of the Israeli Reserve: Present Crises and Future Challenges', *Armed Forces and Society*, 37(2).

Pierce, T. (no date) 'US Naval Institute Proceedings', 122(9).

Porter, P. (2015) *The Global Village Myth*, London: Hurst.

Posen, B. (1984) *The Sources of Military Doctrine: France, Britain and Germany Between the World Wars*, New York: Cornell University Press.

Putnam, R. (2000) *Bowling Alone: The Collapse and Revival of American Community*, London: Simon & Schuster.

Rasmussen, M. (2015) *The Military's Business: Designing Military Power for the Future*, Cambridge: Cambridge University Press.

Reimer, J. (1999) 'A Note from the Chief of Staff of the Army on the Revolution in Military Logistics', *Army Logistician*, Jan./Feb.

Richards, D. (2014) *Taking Command*, London: Headline.

Rosen, S. (1991) *Winning the Next War: Innovation and the Modern Military*, New York: Cornell University Press.

Russell, J. (2010) 'Innovation in War: Counterinsurgency Operations in Anbar and Ninewa Provinces, Iraq, 2005–2007', *Journal of Strategic Studies*, 33(4).

Rutner, S.M., Aviles, M. and Cox, S. (2012) 'Logistics evolution: a comparison of military and commercial logistics thought', *The International Journal of Logistics Management*, 23(1).

Salo, M. (2011) *United We Stand – Divided We Fall: A Standard Model of Unit Cohesion*, Helsinki: Department of Social Research, Helsinki University.

Salo, M. and Siebold, G. (2005) 'Cohesion component as predictors of performance and attitudinal criteria', unpublished paper presented at the Annual Meeting of the International Military Testing Association, Singapore, 7–10 November.

Salo, M. and Sinkko, R. (eds) (2012) *The Science of Unit Cohesion – Its Characteristics and Impacts*, Tampere: Finnish National Defence University.

Sarkesian, S. and Connor, R. (1999) *The US Military Profession in the 21st Century: War, Peace and Politics*, London: Frank Cass.

Satre, L.J. (1976) 'St. John Brodrick and Army Reform 1901–1903', *The Journal of British Studies*, 15(2).

Shils, E. and Janowitz, M. (1948) 'Cohesion and Disintegration in the Wehrmacht in World War II', *Public Opinion Quarterly*, Summer.

Shouesmith, D. (2001) 'Logistics and Support to Expeditionary Operations', *RUSI Defence Systems*, 14(1).

Siebold, G. (1996) 'Small Unit Dynamics: Leadership, Cohesion, Motivation, and Morale', in Phelps, R. and Farr, B. (eds), *Reserve Component Soldiers as Peacekeepers*, Alexandria, VA: U.S. Army Research Institute for the Behavioral and Social Sciences.

—— (1999) 'The Evolution of the Measurement of Cohesion', *Military Psychology*, 11(1).

—— (2007) 'The Essence of Military Cohesion', *Armed Forces and Society*, 33(2).

—— (2012) 'The Science of Military Cohesion', in Salo, M. and Sinkko, R. (eds) *The Science of Unit Cohesion – Its Characteristics and Impacts*, Tampere: Finnish National Defence University.

Siebold, G. and Kelly, D. (1988) *The Development of the Platoon Cohesion Index*, Washington, DC, Army Research Institute.

Simon, S. (2001) 'The Art of Military Logistics', *Communications of the ACM*, 44(6).

Sion, L. and Ben-Ari, E. (2005) 'Hungry, Weary and Horny: Joking and Jesting among Israel's Combat Reserves', *Israel Affairs*, 11(4).

Smith, B. (2007) 'The Mandate to Revolutionize Military Logistics', *Air and Space Power Journal*, 21(2).

Smith, H. and Jans, N. (2011) 'Use Them or Lose Them? Australia's Defence Force Reserves', *Armed Forces and Society*, 37(2).

Smith, Lt. Gen. R. (1995) 'The Commander's Role' in White, Maj. Gen. M. (ed.) *Gulf Logistics: Blackadder's War*, London: Brassey's.

Smith, R. (2005) *The Utility of Force*, London: Allen Lane.

Spiers, E. (1980) *The Army and Society, 1815–1914*, London: Longman.

Spiers, E. (1994) 'The Late Victorian Army 1868–1914', in Chandler, D. and Beckett, I. (eds) *The Oxford History of the British Army*, Oxford: Oxford University Press.

____ (1980) *Haldane: An Army Reformer*, Edinburgh: Edinburgh University Press.

Spradley, J. (1980) *Participant Observation*, Orlando, Florida: Harcourt College Publishers.

Stanhope, H. (1979) *The Soldiers: An Anatomy of the British Army*, London: Hamish Hamilton.

Stouffer, S., Lumsdaine, A., Harper, M., Lumsdaine, R., Williams, J., Brewster Smith, M., Janis, I., Star, S. and Cottrell, L. (1949) *The American Soldier: Combat and Its Aftermath*, Princeton: Princeton University Press.

Strachan, H. (2003) 'The Civil-Military Gap in Britain', *Journal of Strategic Studies*, 26(2).

____ (2006) 'Training, Morale and Modern War', *Journal of Contemporary History*, 41(2).

Swidler, A. (1986) 'Culture in Action: Symbols and Strategies', *American Sociological Review*, 51(2).

Taylor, C. (2010) 'A Brief Guide to Previous British Defence Reviews', London: House of Commons Library.

The War Office (1922) *Statistics of the Military Effort of the British Empire During the Great War 1914–1920*, London: HMSO.

Thornborrow, T. and Brown, A. (2009) 'Being Regimented: Aspiration and Identity Work in the British Parachute Regiment', *Organization Studies*, 30(4).

Thorpe, G. (2012) *Pure Logistics: The Science of War Preparation*, 2nd Edition, Charleston: Nabu Press.

Tilly, C. (1990) *Coercion, Capital, and European States, AD 990–1990*, Cambridge: Blackwell.

Tucker, A. (1963) 'Army and Society in England 1870–1900: A Reassessment of the Cardwell Reforms', *Journal of British Studies*, 2(2).

U.S. Army (2012) *Doctrine Publication 4-0 Sustainment*.

U.S. Department of Defense (2013) *Joint Publication 4-0: Logistics*.

____ (2014) *DoD Supply Chain Materiel Management Procedures: Operational Requirement*.

____ *Defense Reform Initiative Directive No. 54 – Logistics Transformation Plans*.

U.S. Secretary of the Army (2012) *Army Directive 2012–08 Total Force Policy*.

U.S. Senate Committee on Foreign Relations (2011) *Central Asia and the Transition in Afghanistan*, 19 December.

Uttley, M. and Kinsey, C. (2012) 'The Role of Logistics in War', in Strachan, H. (ed.) *The Oxford Handbook of War*, Oxford: Oxford University Press.

Van Creveld, M. (2009) *Supplying War: Logistics from Wallenstein to Patton*, Cambridge: Cambridge University Press.

Vest, B. (2013) 'Citizen, Soldier, or Citizen-Soldier? Negotiating Identity in the U.S. National Guard', *Armed Forces and Society*, 39(4).

Vincent, A. (2007) 'German Philosophy and British Public Policy: Richard Burdon Haldane in Theory and Practice', *Journal of the History of Ideas*, 68(1).

Walker, W. (1990) *Reserve Forces and The British Territorial Army*, London: Tri-Services Press.

—— (1992) 'Comparing Army Reserve Force: A Tale of Multiple Ironies, Conflicting Realities, and More Certain Prospects', *Armed Forces and Society*, 18(3).

Watson, M. and Hay, C. (2003) 'The Discourse of Globalisation and the Logic of No Alternative: Rendering the Contingent Necessary in the Political Economy of New Labour', *Policy and Politics*, 31(3).

Weber, C. (2011) 'The French Military Reserve: Real or Abstract Force?', *Armed Forces and Society*, 37(2).

Wilkinson, S. (1933) *Thirty-Five Years: 1874–1909*, London: Constable.

Williams, R. and Lamb, G. (2010) *Upgrading Our Armed Forces*, London: Policy Exchange.

Winslow, D. (1999) 'Rites of Passage and Group Bonding in the Canadian Airborne', *Armed Forces and Society*, 25(3).

Wong, L. and Gerras, S. (2006) *CU @ The FOB: How the Forward Operating Base is Changing the Life of Combat Soldiers*, Washington, DC: Strategic Studies Institute.

Wong, L., Koldtiz, T., Millem, R. and Potter, T. (2003) *Why They Fight: Combat Motivation in the Iraq War*, Carlisle Barracks, PA: Strategic Studies Institute, U.S. Army War College.

Yoho, K., Rietjens, S. and Tatham, P. (2013) 'Defence Logistics: An Important Research Field in Need of Researchers', *International Journal of Physical Distribution and Logistics Management*, 43(2).

YouGov (2015) *Report on British Attitudes to Defence, Security and the Armed Forces*, available at https://yougov.co.uk/news/2014/10/25/report-british-attitudes-defence-security-and-arme/.

Zurcher, L. (1965) 'The Sailor Aboard Ship: A Study of Role Behaviour in a Total Institution', *Social Forces*, 43(3).

Zurcher, L. and Harries-Jenkins, G. (eds) (1978) *Supplementary Military Forces: Reserves, Militias, Auxiliaries*, Beverly Hills: Sage.

Index